SCHOLAR Study Guide

CfE Advanced Higher F
Unit 3: Electromagnetism
and
Unit 4: Investigating Physics

Authored by:

Unit 3: Julie Boyle (St Columba's School)

Unit 4: Chad Harrison (Tynecastle High School)

Reviewed by:

Grant McAllister (Bell Baxter High School)

Previously authored by:

Andrew Tookey (Heriot-Watt University)

Campbell White (Tynecastle High School)

Heriot-Watt University

Edinburgh EH14 4AS, United Kingdom.

First published 2016 by Heriot-Watt University.

This edition published in 2016 by Heriot-Watt University SCHOLAR.

Copyright © 2016 SCHOLAR Forum.

Distributed by the SCHOLAR Forum.

SCHOLAR Study Guide Unit 3 and Unit 4: CfE Advanced Higher Physics

1. CfE Advanced Higher Physics Course Code: C757 77

     ISBN 978-1-911057-14-7

Printed and bound by CPI Group (UK) Ltd, Croydon, CR0 4YY

Acknowledgements

Thanks are due to the members of Heriot-Watt University's SCHOLAR team who planned and created these materials, and to the many colleagues who reviewed the content.

We would like to acknowledge the assistance of the education authorities, colleges, teachers and students who contributed to the SCHOLAR programme and who evaluated these materials.

Grateful acknowledgement is made for permission to use the following material in the SCHOLAR programme:

The Scottish Qualifications Authority for permission to use Past Papers assessments.

The Scottish Government for financial support.

The content of this Study Guide is aligned to the Scottish Qualifications Authority (SQA) curriculum.

Contents

Topic 1

Electric force and field (Unit 3)

Contents

Prerequisite knowledge

- *Newton's laws of motion.*

- *Gravitational forces and fields (Unit 1 - Topic 5).*

Learning objectives

By the end of this topic you should be able to:

- *carry out calculations involving Coulomb's law for the electrostatic force between point charges $F = \frac{Q_1 Q_2}{4\pi\varepsilon_0 r^2}$;*

- *describe how the concept of an electric field is used to explain the forces that stationary charged particles exert on each other;*

- *draw the electric field pattern around a point charge, a system of charges and in a uniform electric field;*

- *state that the field strength at any point in an electric field is the force per unit positive charge placed at that point in the field, and is measured in units of $N\ C^{-1}$;*

- *perform calculations relating electric field strength to the force on a charged particle $F = QE$;*

- *apply the expression for calculating the electric field strength E at a distance r from a point charge $E = \frac{Q}{4\pi\varepsilon_0 r^2}$;*

- *calculate the strength of the electric field due to a system of charges.*

1.1   Introduction

In Unit 2 you studied the motion of charged particles in a magnetic field. Unit 3 is called Electromagnetism and it will build on this work by studying electric, magnetic and electromagnetic fields. Basic to this work is an understanding of electric charge.

Electric forces act on static and moving electric charges. We will be using the concepts of electric field and electric potential to describe electrostatic interactions.

In this topic we will look at the force that exists between two or more charged bodies, and then introduce the concept of an electric field.

1.2   Electric charge

On an atomic scale, electrical charge is carried by protons (positive charge) and electrons (negative charge). The **Fundamental unit of charge** e is the magnitude of charge carried by one of these particles. The value of e is 1.60×10^{-19} coulombs (C). A charge of one coulomb is therefore equal to the charge on 6.25×10^{18} protons or electrons. It should be noted that one coulomb is an extremely large quantity of charge, and we are unlikely to encounter such a huge quantity of charge inside the laboratory. The sort of quantities of charge we are more likely to be dealing with are of the order of microcoulombs ($1\ \mu C = 10^{-6}$ C), nanocoulombs (1 nC $= 10^{-9}$ C) or picocoulombs (1 pC $= 10^{-12}$ C).

1.3   Coulomb's law

Let us consider two charged particles. We will consider point charges; that is to say, we will neglect the size and shape of the two particles and treat them as two points with charges Q_1 and Q_2 separated by a distance r. The force between the two charges is proportional to the magnitude of each of the charges.
That is to say

$$F \propto Q_1 \quad \text{and} \quad F \propto Q_2$$

The force between the two charges is also inversely proportional to the square of their separation. That is to say

$$F \propto \frac{1}{r^2}$$

These relationships are known as **Coulomb's law**.

The mathematical statement of Coulomb's law is

$$F = \frac{Q_1 Q_2}{4\pi\varepsilon_0 r^2}$$

(1.1)

. . . .

The constant ε_0 is called the permittivity of free space, and has a value of 8.85×10^{-12} F m^{-1} or C^2 N^{-1} m^{-2}. So the constant of proportionality in Equation 1.1 is $\frac{1}{4\pi\varepsilon_0}$, which has the value 8.99×10^9 N m^2 C^{-2}. This force is called the electrostatic or Coulomb force. It is important to remember that force is a vector quantity, and the **direction** of the Coulomb force depends on the sign of the two charges. You should already be familiar with the rule that 'like charges repel, unlike charges attract'. Also, from Newton's third law of motion, we can see that each particle exerts a force of the same magnitude but the opposite direction on the other particle.

Example

Problem:

Two point charges A and B are separated by a distance of 0.200 m. If the charge on A is +2.00 μC and the charge on B is -1.00 μC, calculate the force each charge exerts on the other.

Solution:

Figure 1.1: Coulomb force acting between the two particles

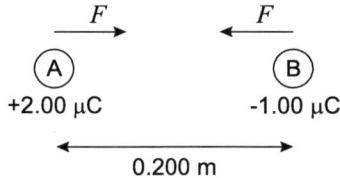

The force acting on the particles is given by Coulomb's law:

. . . .

$$F = \frac{Q_1 \, Q_2}{4\pi\varepsilon_0 r^2}$$

$$\therefore F = \frac{2.00 \times 10^{-6} \times \left(-1.00 \times 10^{-6}\right)}{4\pi \times 8.85 \times 10^{-12} \times (0.200)^2}$$

$$\therefore F = \frac{-2.00 \times 10^{-12}}{4\pi \times 8.85 \times 10^{-12} \times 0.04}$$

$$\therefore F = \frac{-2.00}{4\pi \times 8.85 \times 0.04}$$

$$\therefore F = -0.450 \text{ N}$$

The size of the force is 0.450 N. The minus sign indicates that we have two oppositely charged particles, and hence each charge exerts an attractive force on the other.

. . . .

1.3.1 Electrostatic and gravitational forces

The Coulomb's law equation looks very similar to the equation used to calculate the gravitational force between two particles (Newton's law of gravitation).

$$F = \frac{Q_1 \, Q_2}{4\pi\varepsilon_0 r^2} \qquad\qquad F = G\frac{m_1 m_2}{r^2}$$

$$F \propto \frac{Q_1 \, Q_2}{r^2} \qquad\qquad F \propto \frac{m_1 m_2}{r^2}$$

In both cases, the size of the force follows an **inverse square law** dependence on the distance between the particles. One important difference between these forces is that the gravitational force is always attractive, whereas the direction of the Coulomb force depends on the charge carried by the two particles.

1.3.2 Force between more than two point charges

So far we have used Coulomb's law to calculate the force due to two charged particles, and we find equal and opposite forces exerted on each particle. We will now consider what happens if another charged particle is introduced into the system.

To calculate the force on a charged particle due to two (or more) other charged particles we perform a Coulomb's law calculation to work out each individual force. The **total** force acting on one particle is then the **vector sum** of all the forces acting on it. This makes use of the **principle of superposition of forces**, and holds for any number of charged particles.

Example

Problem:

Earlier we looked at the problem of two particles A (+2.00 μC) and B (-1.00 μC) separated by 0.200 m. Let us now put a third particle X (+3.00 μC) at the midpoint of AB. What is the magnitude of the total force acting on X, and in what direction does it act?

Solution:

Figure 1.2: Separate forces acting on a charge placed between two charged particles

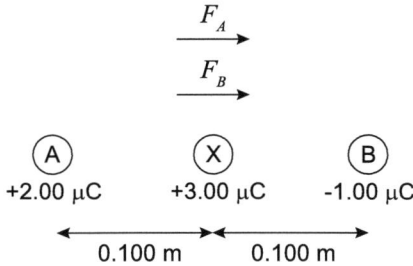

When solving a problem like this, you should always draw a sketch of all the charges, showing their signs and separations. In this case, Figure 1.2 shows the force that A exerts on X is repulsive, and the force that B exerts on X is attractive. Thus both forces act in the **same** direction.

In calculating the size of the two forces we will ignore any minus signs. What we are looking for is just the magnitude of each force. The vector diagram we have drawn shows us the direction of the two forces.

$$F_{AX} = \frac{Q_A Q_X}{4\pi\varepsilon_0 r_{AX}{}^2}$$
$$\therefore F = \frac{2.00 \times 10^{-6} \times 3.00 \times 10^{-6}}{4\pi \times 8.85 \times 10^{-12} \times 0.100^2}$$
$$\therefore F = \frac{6.00 \times 10^{-12}}{4\pi \times 8.85 \times 10^{-12} \times 0.01}$$
$$\therefore F = \frac{6.00}{4\pi \times 8.85 \times 0.01}$$
$$\therefore F = 5.40 \text{ N}$$

$$F_{BX} = \frac{Q_B Q_X}{4\pi\varepsilon_0 r_{BX}{}^2}$$
$$\therefore F = \frac{1.00 \times 10^{-6} \times 3.00 \times 10^{-6}}{4\pi \times 8.85 \times 10^{-12} \times 0.100^2}$$
$$\therefore F = \frac{3.00 \times 10^{-12}}{4\pi \times 8.85 \times 10^{-12} \times 0.01}$$
$$\therefore F = \frac{3.00}{4\pi \times 8.85 \times 0.01}$$
$$\therefore F = 2.70 \text{ N}$$

The vector sum of these two forces is 5.40 + 2.70 = 8.10 N. The direction of the force on X is towards B. The same technique of finding the vector sum would be used in the more general case where the charges were not placed in a straight line.

Three charged particles in a line

Two point charges are separated by 1.00 m. One of the charges (X) is +4.00 μC, the other (Y) is -6.00 μC. A third charge of +1.00 μC is placed between X and Y. Without performing a detailed calculation, sketch a graph to show how the force on the third charge varies as it is moved along the straight line from X to Y.

The total Coulomb force acting on a charged object is equal to the vector sum of the individual forces acting on it.

Quiz: Coulomb force

Go online

Useful data:

Fundamental charge e	1.60×10^{-19} C
Permittivity of free space ε_0	8.85×10^{-12} F m^{-1}

Q1: Two particles J and K, separated by a distance r, carry different positive charges, such that the charge on K is twice as large as the charge on J. If the electrostatic force exerted on J is F N, what is the magnitude of the force exerted on K?

a) $4F$ N
b) $2F$ N
c) F N
d) $F/2$ N
e) $F/4$ N

....

Q2: How many electrons are needed to carry a charge of -1 C?

a) 1.60×10^{-19}
b) 8.85×10^{-12}
c) 1.00
d) 6.25×10^{18}
e) 1.60×10^{19}

....

Q3: A point charge of +5.0 μC sits at point A, 10 cm away from a -2.0 μC charge at point B. What is the Coulomb force acting on the charge placed at point A?

a) 7.2 N away from B.
b) 7.2 N towards B.
c) 9.0 N away from B.
d) 9.0 N towards B.
e) 13.5 N away from B.

....

Q4: A small sphere charged to -4.0 μC experiences an attractive force of 12.0 N due to a nearby point charge of +2.0 μC. What is the separation between the two charged objects?

a) 7.7 cm
b) 6.7 cm
c) 1.0 cm
d) 6.0 mm
e) 4.5 mm

....

Q5: Three point charges X, Y and Z lie on a straight line with Y in the middle, 5.0 cm from both of the other charges. If the values of the charges are $X = +1.0$ μC, $Y = -2.0$ μC and $Z = +3.0$ μC, what is the net force exerted on Y?

a) 0.0 N
b) 14.4 N towards Z
c) 14.4 N towards X
d) 28.8 N towards Z
e) 28.8 N towards X

. . . .

1.4 Electric field strength

Earlier in this topic we noted the similarity between Coulomb's law and Newton's law of gravitation. In our work on gravitation we introduced the idea of a gravitational field. Similarly, an **electric field** can be defined as the space that surrounds electrically charged particles and in which a force is exerted on other electrically charged particles. Just as the gravitational field strength is the force acting per unit mass placed in a gravitational field, electric field strength E is the force F acting per unit positive charge Q placed at a point in the electric field.

$$E = \frac{F}{Q}$$

(1.2)

. . . .

The units of E are N C^{-1}. E is a vector quantity, and the direction of the vector, like the direction of the force F, is the direction of the force acting on a positive charge Q. We can define the electric field strength as being the force acting on a unit positive charge (+1 C) placed in the field. This definition gives us not only the magnitude, but also the direction of the field vector. Rearranging Equation 1.2 as $F = QE$, we can see that this is of the same form as the relationship between gravitational field and force: $F = mg$.

We can calculate the field at a distance r from a point charge Q using Equation 1.1 and Equation 1.2. Starting from Equation 1.1

$$F = \frac{Q_1 \, Q_2}{4\pi\varepsilon_0 r^2}$$

Replacing Q_1 by the charge Q and Q_2 by the unit positive test charge Q_{test} gives us

$$F = \frac{Q \, Q_{test}}{4\pi\varepsilon_0 r^2}$$

Substituting for $F = EQ_{test}$ (from Equation 1.2)

$$EQ_{test} = \frac{Q \, Q_{test}}{4\pi\varepsilon_0 r^2}$$
$$\therefore E = \frac{Q}{4\pi\varepsilon_0 r^2}$$

(1.3)

. . . .

Using this equation, we can calculate the field strength due to a point charge at any position in the field.

Example

Problem:

The electric field strength 2.0 m away from a point charge is 5.0 N C^{-1}. Calculate the value of the charge.

Solution:

To find the charge Q, we need to rearrange Equation 1.3

$$E = \frac{Q}{4\pi\varepsilon_0 r^2}$$
$$\therefore Q = E \times 4\pi\varepsilon_0 r^2$$

Now we can insert the values of E and r given in the question

$$Q = 5.0 \times 4\pi \times 8.85 \times 10^{-12} \times 2.0^2$$
$$\therefore Q = 2.2 \times 10^{-9} \, C$$

. . . .

We can sketch the electrical field lines around the charge as shown in Figure 1.3, but remember that the sign of the charge determines the direction of the field vectors.

Also remember that the closer the field lines, the stronger the electric field. So the spacing of the field lines shows us that the greater the distance from the charge, the weaker the electric field.

Figure 1.3: Field lines around (a) an isolated positive charge; (b) an isolated negative charge

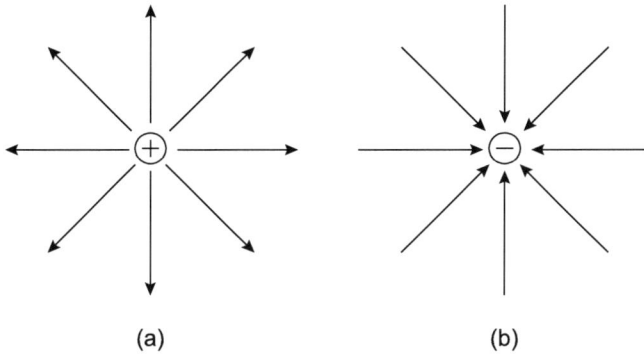

(a)   (b)

. . . .

1.4.1 Electric field due to point charges

The electric field is another vector quantity. We can work out the total electric field due to more than one point charge by calculating the vector sum of the fields of each individual charge.

Example

Problem:

Two point charges Q_A and Q_B are placed at points A and B, where the distance AB = 0.50 m. Q_A = +2 μC. Calculate the total electric field strength at the midpoint of AB if

1. Q_B = +3 μC

2. Q_B = -3 μC.

Solution:

Solution

1. When the two charges have the same sign, the fields at the midpoint of AB are in opposite directions, so the total field strength is given by the difference between them. As shown in Figure 1.4, since Q_A and Q_B are both positive charges, a test charge placed between them will be repulsed by both.

Figure 1.4: Electric fields due to two positive charges

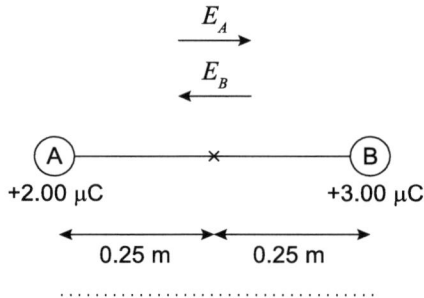

Total field $E = E_B - E_A$

$$\therefore E = \frac{Q_B}{4\pi\varepsilon_0 r^2} - \frac{Q_A}{4\pi\varepsilon_0 r^2}$$

$$\therefore E = \frac{Q_B - Q_A}{4\pi\varepsilon_0 r^2}$$

$$\therefore E = \frac{\left(3.0 \times 10^{-6}\right) - \left(2.0 \times 10^{-6}\right)}{4\pi\varepsilon_0 \left(0.25\right)^2}$$

$$\therefore E = \frac{1.0 \times 10^{-6}}{4\pi\varepsilon_0 \times 0.0625}$$

$$\therefore E = 1.4 \times 10^5 \, \mathrm{N\,C^{-1}}$$

The total field strength is 1.4 × 10⁵ N C⁻¹ directed towards A.

2. When Q_B is a negative charge, the two field components at the midpoint are pointing in the same direction, as shown in Figure 1.5. The total field strength then becomes the sum of the two components.

Figure 1.5: Electric fields due to two opposite charges

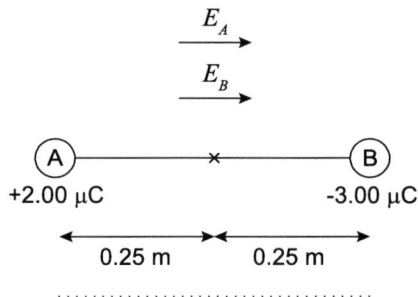

$$E = \frac{Q_A}{4\pi\varepsilon_0 r^2} + \frac{Q_B}{4\pi\varepsilon_0 r^2}$$

$$\therefore E = \frac{Q_A + Q_B}{4\pi\varepsilon_0 r^2}$$

$$\therefore E = \frac{\left(2.0 \times 10^{-6}\right) + \left(3.0 \times 10^{-6}\right)}{4\pi\varepsilon_0 (0.25)^2}$$

$$\therefore E = \frac{5.0 \times 10^{-6}}{4\pi\varepsilon_0 \times 0.0625}$$

$$\therefore E = 7.2 \times 10^5 \, \text{N C}^{-1}$$

The total field strength is 7.2×10^5 N C^{-1} directed towards B. As usual, making a quick sketch showing the charges and their signs helps avoid mistakes.

. . . .

The electric field pattern between two point charges depends upon their polarity. Remember that the field lines always point in the direction that positive charge would move. An electron would move in the opposite direction to the arrows.

Figure 1.6: Electric field pattern for (a) two opposite point charges, (b) two positive point charges and (c) two negative point charges.

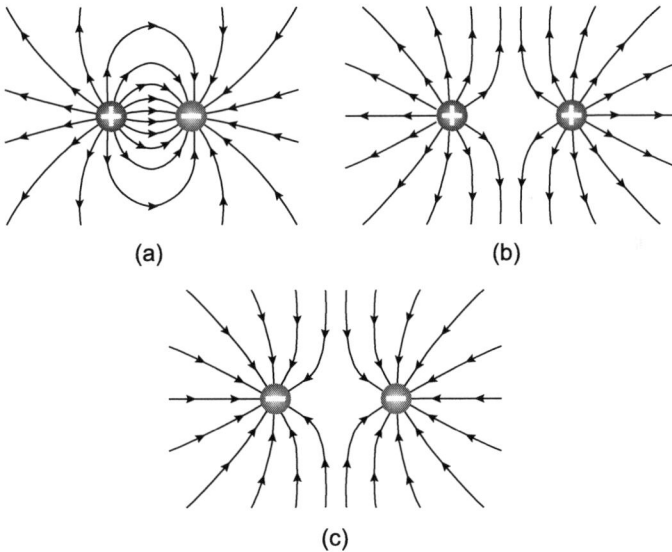

(a)   (b)

(c)

. . . .

Quiz: Electric field

Go online

Useful data:

Fundamental charge e	1.60×10^{-19} C
Permittivity of free space ε_0	8.85×10^{-12} C^2 N^{-1} m^{-2}

Q6: At a distance x m from an isolated point charge, the electric field strength is E N C^{-1}.
What is the strength of the electric field at a distance $2x$ m from the charge?

a) $E/4$
b) $E/2$
c) E
d) $2E$
e) $4E$

. . . .

Q7: A +2.50 nC charged sphere is placed in an electric field of strength 5.00 N C^{-1}.
What is the magnitude of the force exerted on the sphere?

a) 5.00×10^{-10} N
b) 2.00×10^{-9} N
c) 2.50×10^{-9} N
d) 5.00×10^{-9} N
e) 1.25×10^{-8} N

. . . .

Q8: A 50.0 N C^{-1} electric field acts in the positive x-direction.
What is the force on an electron placed in this field?

a) 0.00 N
b) 6.25×10^{-18} N in the -x-direction
c) 6.25×10^{-18} N in the +x-direction
d) 8.00×10^{-18} N in the -x-direction
e) 8.00×10^{-18} N in the +x-direction

. . . .

Q9: Two point charges, of magnitudes +30.0 nC and +50.0 nC, are separated by a distance of 2.00 m.
What is the magnitude of the electric field strength at the midpoint between them?

a) 1.35×10^{-5} N C^{-1}
b) 30.0 N C^{-1}
c) 45.0 N C^{-1}
d) 180 N C^{-1}
e) 720 N C^{-1}

. . . .

Q10: A +1.0 μC charge is placed at point X. A +4.0 μC charge is placed at point Y, 50 cm from X.
How far from X, on the line XY, is the point where the electric field strength is zero?

a)   10 cm
b)   17 cm
c)   25 cm
d)   33 cm
e)   40 cm

. . . .

1.5 Summary

In this topic we have studied electrostatic forces and fields. An electrostatic (Coulomb) force exists between any two charged particles. The magnitude of the force is proportional to the product of the two charges, and inversely proportional to the square of the distance between them. The direction of the force acting on each of the particles is determined by the sign of the charges. If more than two charges are being considered, the total force acting on a particle is the vector sum of the individual forces.

An electric field is a region in which a charged particle will be subject to the Coulomb force. The electric field strength is measured in N C^{-1}. The direction of the electric field vector at a point in a field is the direction in which a Coulomb force would act on a positive charge placed at that point.

Summary

You should now be able to:

- carry out calculations involving Coulomb's law for the electrostatic force between point charges $F = \frac{Q_1 Q_2}{4\pi\varepsilon_0 r^2}$;

- describe how the concept of an electric field is used to explain the forces that stationary charged particles exert on each other;

- draw the electric field pattern around a point charge, a system of charges and in a uniform electric field;

- state that the field strength at any point in an electric field is the force per unit positive charge placed at that point in the field, and is measured in units of N C^{-1} ;

- perform calculations relating electric field strength to the force on a charged particle $F = QE$;

- apply the expression for calculating the electric field strength E at a distance r from a point charge
$$E = \frac{Q}{4\pi\varepsilon_0 r^2} \; ;$$

Summary continued
• calculate the strength of the electric field due to a system of charges.

1.6 Extended information

1.6.1 Electric field around a charged conducting sphere

Let's consider the electric field in the region of the a charged metal sphere. The first point to note is that the charge resides on the surface of the conductor. The electrostatic repulsion between all the individual charges means that, in equilibrium, all the excess charge (positive or negative) rests on the surface. The interior of the conductor is neutral. The distribution of charges across the surface means that **the electric field is zero at any point within a conducting material.**

The same is true of a hollow conductor. Inside the conductor, the field is zero at every point. If we were to place a test charge somewhere inside a hollow charged sphere, there would be no net force acting on it. This fact was first demonstrated by Faraday in his 'ice pail' experiment, and has important applications today in electrostatic screening.

If we plot the electric field inside and outside a hollow conducting sphere, we find it follows a $1/_{r^2}$ dependence outside, but is equal to zero inside, as shown in Figure 1.7

Figure 1.7: Electric field in and around a charged hollow conductor

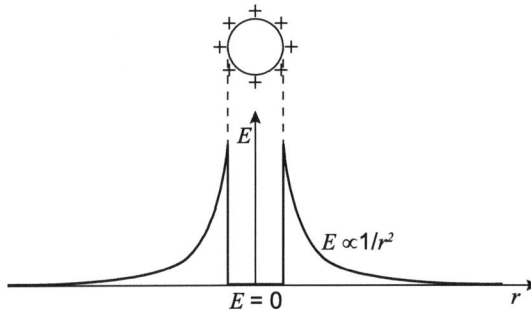

The fact that there is zero net electric field inside a hollow charged conductor means that we can use a hollow conductor as a shield from electric fields. This is another effect that was first demonstrated by Faraday, who constructed a metallic 'cage' which he sat inside holding a sensitive electroscope. As the cage was charged up, there was no deflection of the electroscope, indicating there was no net electric field present inside the cage.

This electrostatic screening is used to protect sensitive electronic circuitry inside equipment such as computers and televisions. By enclosing these parts in a metal box they are shielded from stray electric fields from other appliances (vacuum cleaners etc.). This principle is illustrated in Figure 1.8

Figure 1.8: Electrostatic screening inside a hollow conductor placed in an electric field

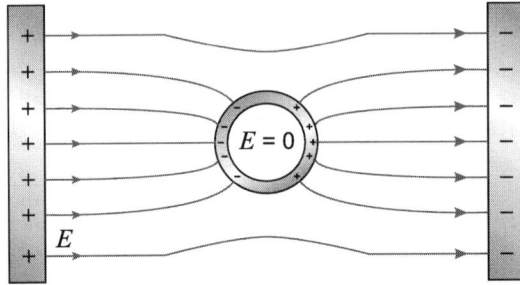

The electric field lines are perpendicular to the surface of the conductor. Induced charge lies on the surface of the sphere, and the net field inside the sphere is zero.

Electrostatic screening also means that the safest place to be when caught in a lightning storm is inside a (metal) car. If the car were to be struck by lightning, the charge would stay on the outside, leaving the occupants safe.

Faraday's ice pail experiment

Go online

There is an online activity which shows how Faraday used a metal ice pail connected to an electroscope to demonstrate that charge resides on the outside of a conductor.

1.6.2 Web links

Web links

There are web links available online exploring the subject further.

1.7 Assessment

End of topic 1 test

Go online

The following test contains questions covering the work from this topic.

The following data should be used when required:

Fundamental charge e	*1.60×10^{-19} C*
Permittivity of free space ε_0	*8.85×10^{-12} F m^{-1}*

Q11: How many electrons are required to charge a neutral body up to -0.032 C?

....

Q12: Two identical particles, each carrying charge +7.65 μC, are placed 0.415 m apart.

Calculate the magnitude of the electrical force acting on each of the particles.

---------- N

....

Q13: The Coulomb force between two point charges is 2.3 N.

Calculate the new magnitude of the Coulomb force if the distance between the two charges is halved.

---------- N

....

Q14: Three charged particles A, B and C are placed in a straight line, with AB = BC = 100 mm. The charges on each of the particles are A = -8.3 μC, B = -2.2 μC and C = +4.3 μC.

What is the magnitude of the total force acting on B?

---------- N

....

Q15: Calculate the electric field strength at a distance 3.2 m from a point charge of +6.5 nC.

---------- N C^{-1}

....

Q16: Two charged particles, one with charge -18 nC and the other with charge +23 nC are placed 1.0 m apart.

Calculate the electric field strength at the point midway between the two particles.

---------- N C^{-1}

....

Q17: A small sphere of mass 0.055 kg, charged to 1.7 μC, is suspended by a thread. If the thread is cut, the sphere will fall to the ground under gravity.

Calculate the magnitude of the electric field acting vertically which would hold the sphere in position when the string is cut.

---------- N C^{-1}

....

Q18: An electron is placed in an electric field of strength 0.013 N C^{-1}.

Calculate the acceleration of the electron.

---------- m s^{-2}

....

Topic 2

Electric potential (Unit 3)

Contents

Prerequisite knowledge

- *Electric force and field (Unit 3 - Topic 1).*

Learning objectives

By the end of this topic you should be able to:

- *state that the electric potential V at a point is the work done by external forces in moving a unit positive charge from infinity to that point;*

- *apply the expression $E = \frac{V}{d}$ for a uniform electric field;*

- *explain what it meant by a conservative field;*

- *state that an electric field is a conservative field;*

- *state and apply the equation $V = \frac{Q}{4\pi\varepsilon_0 r}$ for the potential V at a distance r from a point charge Q.*

2.1   Introduction

In this topic we will be considering the electric potential V. You should already have encountered V in several different contexts - the e.m.f. of a battery, the potential difference across a resistor, calculating the energy stored by a capacitor, and so on. From all these, you should be aware that the potential is a measure of work done or energy in a system. We will be investigating this idea more fully in this topic, and finding how the potential V relates to the electric field E.

We will also investigate the potential due to a point charge and a system of point charges.

2.2   Potential and electric field

In the previous topic, we found similarities between the electric field and force, and the gravitational field and force. We can again draw on this similarity to describe the electric potential.

The gravitational potential tells us how much work is done per unit mass in moving an object which has been placed in a gravitational field. The **electric potential** (or more simply the potential) tells us how much work is done in moving a unit positive charge placed in an electric field. The gravitational potential at a point in a gravitational field was defined as the work done in bringing unit mass from infinity to that point. Similarly, the electric potential V at a point in an electric field can be defined as the work done E_W in bringing unit positive charge Q from infinity to that point in the electric field. This gives us

$$V = \frac{work\,done}{Q}$$
$$\text{or } work\,done = E_W = QV$$

(2.1)

. . . .

From this expression, we can define the unit of electric potential, the volt:

one volt (1 V) = one joule per coulomb (1 J C^{-1})

The **potential difference** V between two points A and B, separated by a distance d, is defined as the work done in moving one unit of positive charge from A to B. Let us consider the case of a uniform electric field E, such as that which exists between the plates of a large parallel-plate capacitor. We will assume A and B lie on the same electric field line.

Figure 2.1: The electric field between the plates of a charged capacitor. The field is uniform everywhere between the plates except near the edges

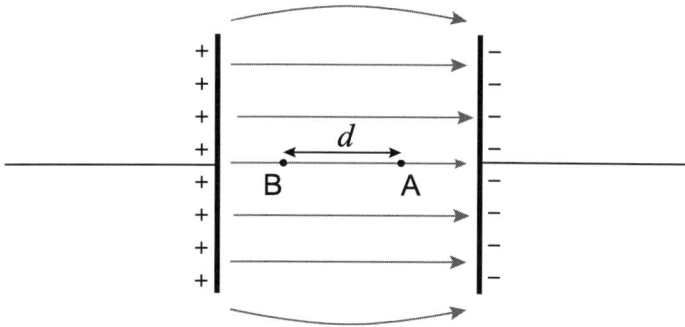

We can use the definition of *work done = force × distance* along with Equation 2.1 to find the work done in moving a unit positive charge from A to B. Remember that the electric field strength is defined as the force acting per unit charge

$$Work\,done = force \times distance$$
$$\therefore QV = (QE) \times d$$
$$\therefore V = E \times d$$

(2.2)

This equation is only valid in the special case of a uniform electric field, where the value of E is constant across the entire distance d.

Example

Problem:

The potential difference between the two plates of a charged parallel-plate capacitor is 12 V. What is the electric field strength between the plates if their separation is 200 μm?

Solution:

Rearranging Equation 2.2 gives us

$$E = \frac{V}{d}$$
$$\therefore E = \frac{12}{200 \times 10^{-6}}$$
$$\therefore E = 6.0 \times 10^4 \, \text{N} \, \text{C}^{-1}$$

. . . .

Looking at the rearranged Equation 2.2, the electric field strength is given as a potential difference divided by a distance. This means that we can express E in the units V m^{-1}, which are equivalent to N C^{-1}.

There is one final point we should note about potential difference. Looking back to Figure 2.1, we moved a unit charge directly from A to B by the shortest possible route. The law of conservation of energy tells us that the work done in moving from A to B is independent of the route taken.

Figure 2.2: Different routes from A to B

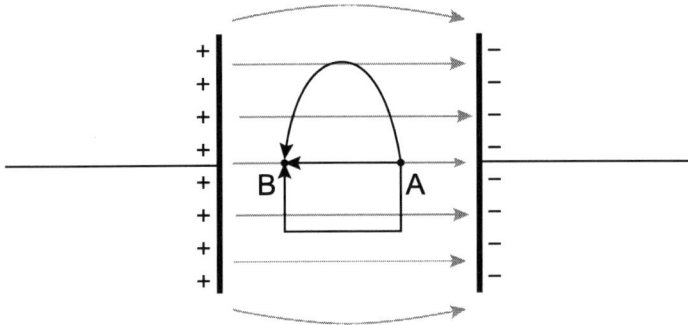

. . . .

Irrespective of the route taken, the start and finish points are the same in Figure 2.2. If the potential difference is V between A and B, the same amount of work must be done in moving a unit of charge from A to B, whatever path is taken. This is because the electric field is a **conservative field**.

Quiz: Potential and electric field

Go online

Useful data:

Fundamental charge e	1.60×10^{-19} C
Permittivity of free space ε_0	8.85×10^{-12} F m^{-1}

Q1: The uniform electric field between two plates of a charged parallel-plate capacitor is 4000 N C^{-1}. If the separation of the plates is 2.00 mm, what is the potential difference between the plates?

a) 500 mV
b) 2.00 V
c) 8.00 V
d) 2000 V
e) 8000 V

. . . .

Q2: Which of the following corresponds to the units of electric field?

a) $J\,C^{-1}$
b) $N\,m^{-1}$
c) $J\,V^{-1}$
d) $V\,m^{-1}$
e) $N\,V^{-1}$

....

Q3: A particle carrying charge + 20 mC is moved through a potential difference of 12 V. How much work is done on the particle?

a) 0.24 J
b) 0.60 J
c) 1.67 J
d) 240 J
e) 600 J

....

Q4: If 4.0 J of work are done in moving a 500 μC charge from point M to point N, what is the potential difference between M and N?

a) 2.0×10^{-5} V
b) 0.80 V
c) 12.5 V
d) 2000 V
e) 8000 V

....

2.3   Electric potential around a point charge and a system of charges

2.3.1   Calculating the potential due to one or more charges

Let us consider a positive point charge Q, and the potential at a distance r from the charge.

Use r to represent distance:

$$E = -\frac{dV}{dr}$$

We also know that the electric field E at a distance r from a point charge Q is given by

$$E = \frac{Q}{4\pi\varepsilon_0 r^2}$$

Combining these, we obtain

$$-\frac{dV}{dr} = \frac{Q}{4\pi\varepsilon_0 r^2}$$

We can integrate this expression to enable us to determine the potential V at a distance r from the point charge. Remembering that the potential due to a point charge is zero at an infinite distance from the charge, the limits for the integration will be $(V = 0, x = \infty)$ and $(V = V, x = r)$

$$-\int_0^V dV = \int_\infty^r \frac{Q}{4\pi\varepsilon_0 r^2} dr$$
$$\therefore -\int_0^V dV = \frac{Q}{4\pi\varepsilon_0} \int_\infty^r \frac{1}{r^2} dr$$
$$\therefore -[V]_0^V = \frac{Q}{4\pi\varepsilon_0} \left[\frac{-1}{r}\right]_\infty^r$$
$$\therefore -V = \frac{Q}{4\pi\varepsilon_0} \left(\frac{-1}{r} + 0\right)$$
$$\therefore V = \frac{Q}{4\pi\varepsilon_0 r}$$

We need to be careful with the sign of the potential. In moving a positive charge from infinity to r, the charge will have gained potential energy, as work has to be done on the charge against the electric field. So if we have defined the potential to be zero at infinity, the potential V must be positive for all r less than infinity. Thus the potential at r is given by

$$V = \frac{Q}{4\pi\varepsilon_0 r}$$

(2.3)

....

Note the difference here between electric and gravitational potential. In both cases, we define zero potential at $r = \infty$. The difference is that in moving unit mass from infinity in a gravitational field, the field does work on the mass, making the potential less than at infinity, and hence a negative number. In moving a unit **positive** charge, we must do work **against** the E-field, so the potential increases, and hence is a positive number.

Unlike the electric field, the electric potential around a point charge decays as $1/r$, not $1/r^2$. The potential is a scalar quantity, not a vector quantity, although its sign is determined by the sign of the charge Q. The field strength and potential around a positive point charge are plotted in Figure 2.3

Figure 2.3: Plots of field strength and potential with increasing distance from a point charge

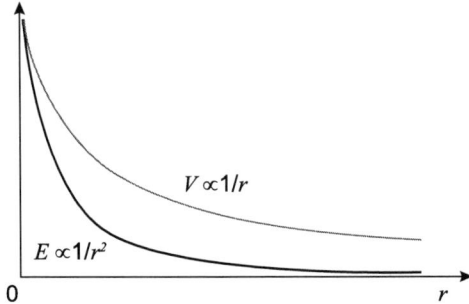

Example

Problem:

At a distance 40 cm from a positive point charge, the electric field is 200 N C^{-1} and the potential is 24 V. What are the electric field strength and electric potential 20 cm from the charge?

Solution:

The electric field strength E varies as $1/_{r^2}$, so if the distance from the charge is halved, the field strength increases by 2^2.

That is to say

$$E_r \propto \frac{1}{r^2}$$

So if we replace r by $r/_2$ in the above equation

$$E_{r/2} \propto \frac{1}{(r/2)^2} = \frac{4}{r^2}$$

So

$$\frac{E_{r/2}}{E_r} = \frac{4/_{r^2}}{1/_{r^2}} = 4$$

E therefore increases by a factor of four, and the new value of E is $4 \times 200 = 800$ N C^{-1}.

V is proportional to $1/_r$, so if r is halved, V will double. Hence the new value of V is 48 V.

When more than one charged particle is present, we can calculate the total potential at a point by adding the individual potentials. This is similar to the way in which we worked out the total Coulomb force and the total electric field. Once again, care must be taken with the signs of the different charges present.

Example

Problem:

What is the net electric potential at the point midway between two point charges of +2.00 μC and -5.00 μC, if the two charges are 2.00 m apart?

Solution:

Figure 2.4: Two charged objects 2.00 m apart

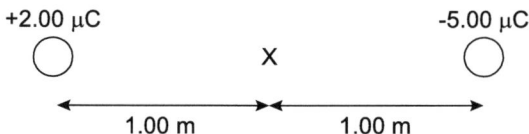

+2.00 μC   -5.00 μC

X

1.00 m   1.00 m

As usual, a sketch such as Figure 2.4 helps in solving the problem.

The potential due to the positive charge on the left is

$$V_1 = \frac{Q_1}{4\pi\varepsilon_0 r_1}$$
$$\therefore V_1 = \frac{2.00 \times 10^{-6}}{4\pi\varepsilon_0 \times 1.00}$$
$$\therefore V_1 = 1.80 \times 10^4 \text{ V}$$

The potential due to the negative charge on the right is

$$V_2 = \frac{Q_2}{4\pi\varepsilon_0 r_2}$$
$$\therefore V_2 = \frac{-5.00 \times 10^{-6}}{4\pi\varepsilon_0 \times 1.00}$$
$$\therefore V_2 = -4.50 \times 10^4 \text{ V}$$

Combining these, the total potential at the mid-point is

$$V = V_1 + V_2$$
$$\therefore V = \left(1.80 \times 10^4\right) - \left(4.50 \times 10^4\right)$$
$$\therefore V = -2.70 \times 10^4 \text{ V}$$

Quiz: Electrical potential due to point charges

Useful data:

Fundamental charge e	*1.60 × 10^{-19} C*
Permittivity of free space ε_0	*8.85 × 10^{-12} F m^{-1}*

Q5: Calculate the electrical potential at a distance of 250 mm from a point charge of +4.0 μC.

a) 0.58 V
b) 2.3 V
c) 1.4 × 10^5 V
d) 5.8 × 10^5 V
e) 1.8 × 10^6 V

. . . .

Q6: An alpha particle has a charge of 3.2 × 10^{-19} C.

Determine the potential energy of an alpha particle at the position outlined in the previous question.

a) -4.48 × 10^{-14} J
b) 1.14 × 10^{-24} J
c) 2.29 × 10^{-24} J
d) 2.24 × 10^{-14} J
e) 4.48 × 10^{-14} J

. . . .

Q7:

+4.0 μC   -5.0 μC

X

←——————— 1.00 m ———————→←——— 0.50 m ———→

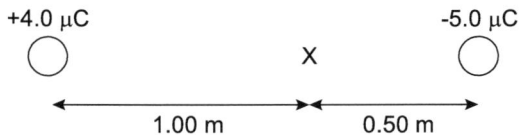

Determine the electrical potential at position X.

a) -7.2 × 10^4 V
b) -1.8 × 10^5 V
c) -1.4 × 10^5 V
d) 7.2 × 10^4 V
e) 1.4 × 10^5 V

. . . .

Q8: At a point 20 cm from a charged object, the ratio of electric field strength to electric potential (E/V) equals 100 m⁻¹.

What is the value of E/V 40 cm from the charge?

a) 25 m⁻¹
b) 50 m⁻¹
c) 100 m⁻¹
d) 200 m⁻¹
e) 400 m⁻¹

. . . .

2.4 Summary

We have defined electric potential and considered the potential difference between two points. We have also shown how electric field and electric potential are related.

Summary

You should now be able to:

- state that the electric potential V at a point is the work done by external forces in moving a unit positive charge from infinity to that point;

- apply the expression $E = \frac{V}{d}$ for a uniform electric field;

- explain what it meant by a conservative field;

- state that an electric field is a conservative field;

- state and apply the equation $V = \frac{Q}{4\pi\varepsilon_0 r}$ for the potential V at a distance r from a point charge Q.

2.5 Extended information

2.5.1 Potential around a hollow conductor

In the previous topic we plotted the electric field in and around a hollow conductor. We found that the field followed a $1/_{r^2}$ dependence outside the conductor, but was equal to zero on the inside. What happens to the potential V inside a hollow conducting shape?

The example shown in Figure 2.5 is a hollow sphere. The definition of potential at any point is the work done in moving unit charge from infinity to that point. So outside the sphere, the potential follows the $1/_r$ dependence we have just derived. Inside the sphere, the total field is zero, so there is no additional work done in moving charge about inside the sphere. The potential is therefore constant inside the sphere, and has the same value as at the edge of the sphere.

Figure 2.5: Electric potential in and around a hollow conducting sphere

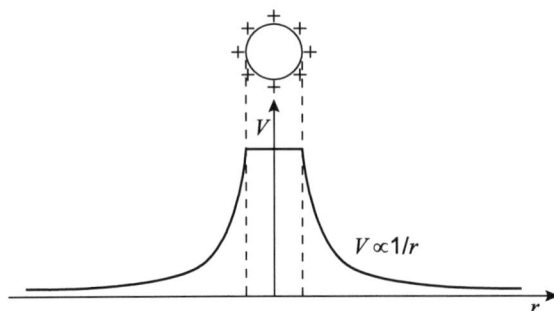

2.5.2 Web links

Web links

There are web links available online exploring the subject further.

2.6 Assessment

End of topic 2 test

The following test contains questions covering the work from this topic.

Go online

The following data should be used when required:

Fundamental charge e	1.60×10^{-19} C
Permittivity of free space ε_0	8.85×10^{-12} F m^{-1}

Q9:   1.7 mJ of work are done in moving a 6.3 μC charge from point A to point B.

Calculate the potential difference between A and B.

_ _ _ _ _ _ _ _ _ _ V

Q10: A 8.8 mC charged particle is moved at constant speed through a potential difference of 7.2 V.

Calculate how much work is done on the particle.

_ _ _ _ _ _ _ _ _ _ J

Q11: The potential difference between points M and N is 80 V and the uniform electric field between them is 1600 N C^{-1}.

Calculate the distance between M and N.

_____ m

. . . .

Q12: A charged particle is moved along a direct straight path from point C to point D, 28 mm away in a uniform electric field. 26 J of work are done on the particle in moving it along this path.

How much work must be done in order to move the particle from C to D along an indirect, curved path of distance 56 mm?

_____ J

. . . .

Q13: Calculate the electric potential at a distance 0.62 m from a point charge of 26 μC.

_____ V

. . . .

Q14: Two point charges, one of +2.5 nC and the other of +3.7 nC are placed 1.9 m apart.

Calculate the electric potential at the point midway between the two charges.

_____ V

. . . .

Q15: A +3.25 μC charge X is placed 2.50 m from a -2.62 μC charge Y.

Calculate the electric potential at the point 1.00 m from X, on the line joining X and Y.

_____ V

. . . .

Q16: At a certain distance from a point charge, the electric field strength E is 2800 N C^{-1} and the electric potential V is 6300 V.

1. Calculate the distance from the charge at which E and V are being measured.
   _____ m
2. Calculate the magnitude of the charge.
   _____ C

. . . .

Topic 3

Motion in an electric field (Unit 3)

Contents

Prerequisite knowledge

- *Coulomb's law (Unit 3 - Topic 1).*

- *Electric potential and the volt (Unit 3 - Topic 2).*

- *Kinematic relationships (Unit 1 - Topic 1).*

- *Newton's laws of motion.*

Learning objectives

By the end of this topic you should be able to:

- *describe the energy transformation that takes place when a charged particle is moving in an electric field;*

- *carry out calculations using $E_w = QV$;*

- *define an electronvolt;*

- *describe the motion of a charged particle in a uniform electric field, and use the kinematic relationships to calculate the trajectory of this motion;*

- *perform calculations to solve problems involving charged particles in electric fields, including the collision of a charged particle with a stationary nucleus.*

3.1   Introduction

In this topic we will study the motion of charged particles in an electric field, drawing upon some of the concepts of the previous two topics. As we progress through this topic, we will also come across some other concepts you should have met before: Newton's second law of motion and Rutherford scattering of α-particles. You may find it useful to refresh your memory of these subjects before starting this topic.

As well as studying the theory, we will also be looking at some of the practical applications of applying electric fields to moving charged particles, such as the cathode ray tubes found in oscilloscopes.

3.2   Energy transformation associated with movement of charge

We have already defined one volt as being equivalent to one joule per coulomb. That is to say, if a charged particle moves through a potential difference of 1 V, it will gain or lose 1 J of energy per coulomb of its charge. Put succinctly, the energy E_W gained by a particle of charge Q being accelerated through a potential V is given by

$$E_W = QV$$

(3.1)

. . . .

A charged particle placed in an electric field will be acted on by the Coulomb force. If it is free to move, the force will accelerate the particle, hence it will gain kinetic energy. So we can state, using Equation 3.1, that the gain in kinetic energy is

$$\frac{1}{2}mv^2 = QV$$

(3.2)

. . . .

Hence we can calculate the velocity gained by a charged particle accelerated by a potential. Note that this is the velocity gained **in the direction of the electric field vector**. You should remember from the previous topic that the electric field points from high to low potential.

Example

Problem:

An electron is accelerated from rest through a potential of 50 V. What is its final velocity?

Solution:

In this case, Equation 3.2 becomes

$$\frac{1}{2}m_e v^2 = eV$$

Rearranging this equation

$$v = \sqrt{\frac{2eV}{m_e}}$$

Now we use the values of $e = 1.60 \times 10^{-19}$ C and $m_e = 9.11 \times 10^{-31}$ kg in this equation

$$v = \sqrt{\frac{2 \times 1.60 \times 10^{-19} \times 50}{9.11 \times 10^{-31}}}$$
$$\therefore v = 4.19 \times 10^6 \, \mathrm{m\,s^{-1}}$$

. . . .

In general, we can state that for a particle of charge Q and mass m, accelerated from rest through a potential V, its final velocity will be

$$v = \sqrt{\frac{2QV}{m}}$$

(3.3)

. . . .

An electron accelerated from rest can attain an extremely high velocity from acceleration through a modest electric potential. It should be noted that as the velocity of the electron increases, or the accelerating potential is increased, relativistic effects will become significant. Broadly speaking, once a charged particle is moving with a velocity greater than 10% of the speed of light ($c = 3.00 \times 10^8$ m s⁻¹) then relativistic effects need to be taken into account.

Quiz: Acceleration and energy change

Go online

Useful data:

Fundamental charge e	1.6×10^{-19} C
Mass of an electron m_e	9.11×10^{-31} kg
Speed of light c	3.00×10^8 m s^{-1}
Mass of an α-particle	6.65×10^{-27} kg
Charge of an α-particle	$+3.20 \times 10^{-19}$ C

Q1: A free electron is accelerated towards a fixed positive charge.

Which **one** of the following statements is true?

a) The electron gains kinetic energy.
b) The electron loses kinetic energy.
c) There are no force acting on the electron.
d) The electron's velocity is constant.
e) A repulsive force acts on the electron.

. . . .

Q2: A -3.0 μC charge is accelerated through a potential of 40 V.

How much energy does it gain?

a) 7.5×10^{-8} J
b) 1.2×10^{-4} J
c) 0.075 J
d) 13 J
e) 120 J

. . . .

Q3: An α-particle is accelerated from rest through a potential of 1.00 kV.

What is its final velocity?

a) 9.80×10^3 m s^{-1}
b) 2.19×10^5 m s^{-1}
c) 3.10×10^5 m s^{-1}
d) 3.00×10^8 m s^{-1}
e) 9.62×10^{10} m s^{-1}

. . . .

Q4: Two parallel plates are 50 mm apart. The electric field strength between the plates is 1.2×10^4 N C^{-1}.

An electron is accelerated between the plates. How much kinetic energy does it gain?

a) 3.8×10^{-17} J
b) 9.6×10^{-17} J
c) 1.9×10^{-15} J
d) 3.8×10^{-14} J
e) 9.6×10^{-14} J

. . . .

3.3 Motion of charged particles in uniform electric fields

If a charged particle is placed in an electric field, we know that the force acting on it will be equal to QE, where Q is the charge on the particle and E is the magnitude of the electric field. If this is the only force acting on the particle, and the particle is free to move, then Newton's second law of motion tells us that it will be accelerated. If the particle has mass m, then

$$
\begin{aligned}
F &= ma \\
\therefore QE &= ma \\
\therefore a &= \frac{QE}{m}
\end{aligned}
$$

(3.4)

. . . .

Equation 3.4 tells us the magnitude of the particle's acceleration **in the direction of the electric field**. We must be careful with the direction of the acceleration. The E-field is defined as positive in the direction of the force acting on a positive charge. An electron will be accelerated in the opposite direction to the E-field vector, whilst a positively-charged particle will be accelerated in the same direction as the E-field vector.

Let us consider what happens when an electron enters a uniform electric field at right angles to the field, as shown in Figure 3.1.

Figure 3.1: An electron travelling through an electric field

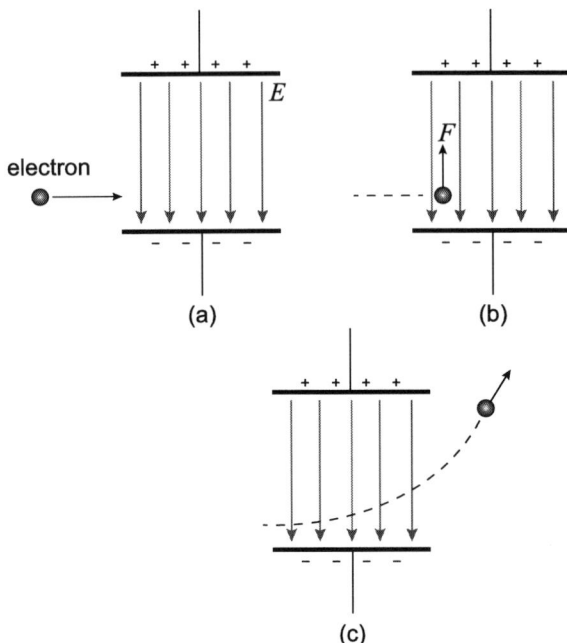

(a)

(b)

(c)

Note that the motion of the electron is similar to that of a body projected horizontally in the Earth's gravitational field. At right angles (orthogonal) to the field, there is no force acting and the electron moves with a uniform velocity in this direction. Parallel to the field, a constant force ($F = QE$, analogous to $F = mg$) acts causing a uniform acceleration parallel to the field lines. This means that we can solve problems involving charged particles moving in electric fields in the same way that we solved two-dimensional trajectory problems, splitting the motion into orthogonal components and applying the kinematic relationships of motion with uniform acceleration.

Example

Problem:

Two horizontal plates are charged such that a uniform electric field of strength E = 200 N C^{-1} exists between them, acting upwards. An electron travelling horizontally enters the field with speed 4.00×10^6 m s^{-1}, as shown in Figure 3.2.

1. Calculate the acceleration of the electron.

2. How far (vertically) is the electron deflected from its original path when it emerges from the plates, given the length l = 0.100 m?

Figure 3.2: An electron travelling in an electric field

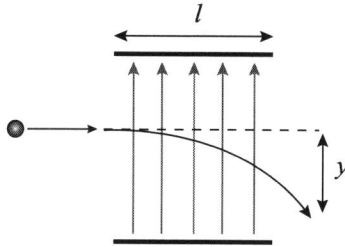

Solution:

1. Considering the vertical motion, we use Equation 3.4 to calculate the downward acceleration

$$a = \frac{QE}{m}$$
$$\therefore a = \frac{eE}{m_e}$$
$$\therefore a = \frac{1.60 \times 10^{-19} \times 200}{9.11 \times 10^{-31}}$$
$$\therefore a = 3.51 \times 10^{13} \text{ m s}^{-2}$$

Note that this acceleration is many orders of magnitude greater than the acceleration due to gravity acting on the electron. We will be able to ignore the effects of gravity in all the problems we encounter concerning the motion of charged particles in electric fields since the effects of the E-fields will always be far greater.

2. To calculate the deflection y, we first need to calculate the time-of-flight of the electron between the plates, which we do by considering the horizontal motion of the electron. Since this is unaffected by the E-field, the horizontal component of the velocity is unchanged and we can use the simple relationship

$$t = \frac{l}{v_h}$$
$$\therefore t = \frac{0.100}{4.00 \times 10^6}$$
$$\therefore t = 2.50 \times 10^{-8} \text{ s}$$

Now, considering the vertical component of the motion, we know $u_v = 0$ m s^{-1}, $t = 2.50 \times 10^{-8}$ s and $a = 3.51 \times 10^{13}$ m s^{-2}. The displacement y is the unknown, so we will use the kinematic relationship $s = ut + \frac{1}{2}at^2$.

In this case

$$y = u_v t + \tfrac{1}{2}at^2$$

$$\therefore y = 0 + \left(\tfrac{1}{2} \times 3.51 \times 10^{13} \times \left(2.50 \times 10^{-8}\right)^2\right)$$

$$\therefore y = 0.0110 \text{ m}$$

Note that we would normally measure the potential difference across the plates. If the plates are separated by a distance d, then the electric field $E = V/d$.

. . . .

Quiz: Charged particles moving in electric fields

Go online

Useful data:

Charge on electron e	*-1.60 x 10^{-19} C*
Mass of an electron m_e	*9.11 × 10^{-31} kg*
Speed of light in vacuum c	*3.00 × 10^8 m s^{-1}*
Mass of an α-particle m_p	*6.65 × 10^{-27} kg*
Charge of an α-particle	*+3.20 × 10^{-19} C*

Q5: An electron enters a region where the electric field strength is 2500 N C^{-1}. What is the force acting on the electron?

a) 6.40 × 10^{-23} N
b) 1.60 × 10^{-19} N
c) 4.00 × 10^{-16} N
d) 2500 N
e) 1.56 × 10^{19} N

. . . .

Q6: An electron is placed in a uniform electric field of strength 4.00 × 10^3 N C^{-1}. What is the acceleration of the electron?

a) 4.00 × 10^{-23} m s^{-2}
b) 1.42 × 10^{-15} m s^{-2}
c) 4.00 × 10^3 m s^{-2}
d) 7.03 × 10^{14} m s^{-2}
e) 2.50 × 10^{22} m s^{-2}

. . . .

Q7: A positively-charged ion placed in a uniform electric field will be

a) accelerated in the direction of the electric field.
b) accelerated in the opposite direction to the electric field.
c) moving in a circular path.
d) moving with constant velocity.
e) stationary.

....

Q8: An α-particle enters a uniform electric field of strength 50.0 N C^{-1}, acting vertically downwards.

What is the acceleration of the particle?

a) 1.20×10^9 m s^{-2} downwards
b) 1.20×10^9 m s^{-2} upwards
c) 2.41×10^9 m s^{-2} downwards
d) 2.41×10^9 m s^{-2} upwards
e) 8.78×10^{12} m s^{-2} downwards

....

Q9: An electron moving horizontally at 1800 m s^{-1} enters a vertical electric field of field strength 1000 N C^{-1}. The electron takes 2.00×10^{-8} s to cross the field.

With what vertical component of velocity does it emerge from the field?

a) 0.00 m s^{-1}
b) 3.51×10^{-2} m s^{-1}
c) 1.80×10^3 m s^{-1}
d) 3.51×10^6 m s^{-1}
e) 7.04×10^6 m s^{-1}

....

3.4 Applications of charged particles and electric fields

We have discussed how electrons can be accelerated and deflected in an electric field. We now look at some practical applications of these effects.

3.4.1 Cathode ray tubes

Cathode ray tubes used to very common. Televisions, computer monitors and oscilloscopes up to about the year 2000 were nearly always made using a cathode ray tube. This meant that these devices were very large. The advent of LCD, LED and plasma screens means that cathode ray tubes have nearly all disappeared from people's homes. However the cathode ray tube is still valuable as a tool for studying electric fields and as an introduction to particle accelerators.

In a cathode ray tube, such as the one shown in Figure 3.3, electrons ('cathode rays') are freed from the heated cathode. (The electrons were originally called cathode rays because these experiments were first carried out before the electron was discovered. To the original experimenters it looked like the cathode was emitting energy rays.) These electrons are accelerated while in the electric field set up between the cathode and the anode, gaining kinetic energy. Some electrons pass through a hole in the anode. From the anode to the y-plates, the electrons travel in a straight line at constant speed, obeying Newton's first law of motion.

Figure 3.3: The cathode ray tube

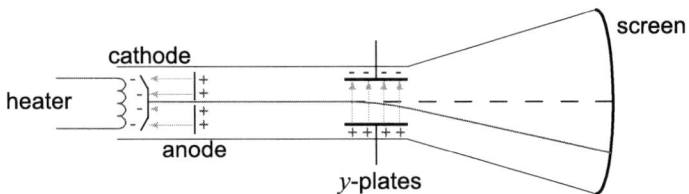

A second electric field is set up between the y-plates, this time at right angles to the initial direction of motion of the electrons. This electric field supplies a force to the electrons at right angles to their original direction. The resulting path of the electrons is a parabola. The motion of the electrons while between the y-plates is similar to the motion of a projectile thrown horizontally in a gravitational field.

When they leave the region of the y-plates, the electrons again travel in a straight line with constant speed (now in a different direction), eventually hitting the screen as shown.

The point on the screen where the electrons hit is determined by the strength of the electric field between the y-plates. This electric field strength is in turn determined by the potential difference between the y-plates. So the deflection of the electron beam can be used to measure a potential difference.

Example

Problem:

The potential between the cathode and the anode of a cathode ray tube is 200 V.

Assuming that the electrons are given off from the heated cathode with zero velocity and that all of the electrical energy given to the electrons is transformed to kinetic energy, calculate

1. the electrical energy gained by an electron between the cathode and the anode.

2. the horizontal velocity of an electron just as it leaves the anode.

(The mass of an electron is 9.11×10^{-31} kg)

Solution:

1. The electrical energy gained by an electron is equal to the work done by the electric field between the cathode and the anode, so

$E_W = QV$
$E_W = 1.6 \times 10^{-19} \times 200$
$E_W = 3.2 \times 10^{-17} J$

2. If all of this energy is transformed to kinetic energy, then

$$E_k = \frac{1}{2}mv^2$$

$$3.2 \times 10^{-17} = \frac{1}{2} \times 9.11 \times 10^{-31} \times v^2$$

$$v = 8.4 \times 10^6 \text{m s}^{-1}$$

....

The cathode ray tube

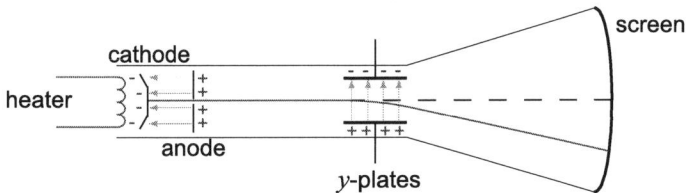

This activity allows you to see the path of electrons in the electric field set up between the cathode and the anode in a cathode ray tube, and calculate the kinetic energy gained by an electron. It also allows the path of the electrons to be changed by applying a potential difference between the y-plates.

Electrons given off from a heated cathode in a cathode ray tube are accelerated by the electric field set up between the cathode and the anode.

The path of the electrons can be changed by the electric field set up by applying a potential difference between the y-plates.

It is important to realise that increasing the potential difference between the cathode and the anode increases the speed of the electrons in the cathode ray. Altering the potential difference between the y-plates affects the position where the electrons hit the screen.

....

3.4.2 Particle accelerators

Particle accelerators are tools that are used to prise apart the nuclei of atoms and thereby help us increase our understanding of the nature of matter and the rules governing the particles and their interaction in the sub atomic world. Particle accelerators are massive machines that accelerate charged particles (ions) and give them enough energy to separate the constituent particles of the nucleus. They have played a significant part in the development of the standard model.

We have already met a very simple particle accelerator: the cathode ray tube. In a cathode ray tube electrons are accelerated by an electric field.

The cathode ray tube however cannot produce high enough energies to investigate the structure of matter. Larger and much more powerful particles have been developed for this purpose.

Particle accelerators are of two main types.

One type accelerates the particle in a straight line. This type is called a linear accelerator, sometimes referred to as a "linac".

The other type accelerates the particle in a circular path. The cyclotron and the more widely used synchrotron are examples of this type.

The linear and circular accelerators both use electric fields as the means of accelerating particles and supplying them with energy.

Linear accelerator

In a linear accelerator, the particle acquires energy in a similar way to the electron in the cathode ray tube but the process is repeated a large number of times. A large alternating voltage is used to accelerate particles along in a straight line.

Figure 3.4: A linear accelerator

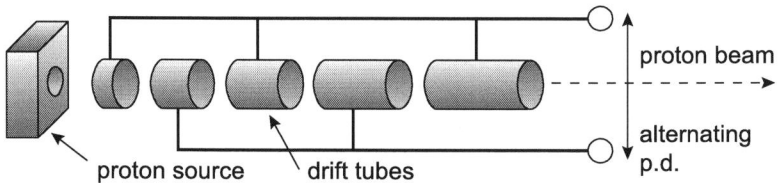

The particles pass through a line of hollow metal tubes enclosed in a long evacuated cylinder. The frequency of the alternating voltage is set so that the particle is accelerated forward each time it goes through a gap between two of the metal tubes. The metal tubes are known as drift tubes. The idea is that the particle drifts free of electric fields through these tubes at constant velocity and emerges from the end of a tube just in time for the alternating voltage to have changed polarity. The largest linac in the world, at Stanford University in the USA, is 3.2 km long.

At the end of each drift tube the charged particle is accelerated by the voltage across the gap.

- The work done on the charged particle, $E_W = QV$
- Where V = voltage across gap, Q = charge on particle being accelerated.
- The particle gains QV of energy at each gap
- This work done causes the particle to accelerate
- So the E_k increases by QV at each gap.
- The speed increases as it moves along the linear accelerator.

The length of successive drift tubes increases. This is because the speed of the charged particle is increasing and to ensure that the time taken to pass through each tube is the same, the length of the tubes must be increased. The time to pass through each drift tube is set by the frequency of the alternating voltage.

It would appear that longer linear accelerators, if they were to be built, could produce particles with yet higher speeds and energy. However special relativity sets limits on the speeds that can be achieved. At speeds comparable with the speed of light (relativistic speeds), the mass of a particle increases significantly and consequently much more energy is needed to accelerate the particle.

Linear accelerators work well but they are expensive and need a lot of space.

3.4.3   Rutherford scattering

In this famous experiment, first carried out by Rutherford and his students Geiger and Marsden in 1909, a stream of alpha particles is fired at a thin sheet of gold foil. Rutherford found that although most particles travelled straight through the foil, a few were deflected, sometimes through large angles, as shown in Figure 3.5(a). Some were even deflected straight back in the direction they had come from. From this experiment, Rutherford concluded that atoms were mostly empty space, with a dense positively-charged nucleus at the centre. The scattering of α-particles was due to collisions with these nuclei.

We can now look a little closer at what happens in these 'collisions'. We can see in Figure 3.5(b) that the α-particle doesn't actually impinge upon the nucleus. Instead there is an electrostatic repulsion between the α-particle and the nucleus. It is the kinetic energy of the α-particle that determines how close it can get to the nucleus.

Figure 3.5: Rutherford scattering

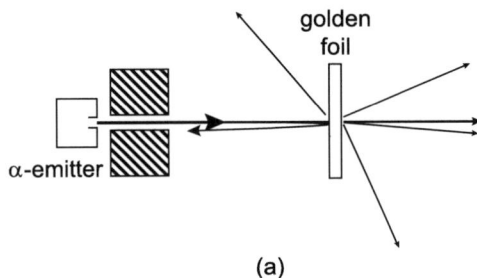

(a)

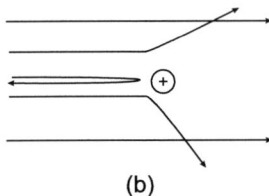

(b)

. . . .

Example

Problem:

In a Rutherford scattering experiment, a beam of alpha particles is fired at a sheet of gold foil. Each α-particle has charge $2e$ and (non-relativistic) energy $E_\alpha = 1.00 \times 10^{-13}$ J. A gold nucleus has charge $79e$. What is the closest possible distance an α-particle with this energy can get to a gold nucleus?

Solution:

As the α-particle approaches the nucleus, its potential energy increases since it is moving in the electric field of the nucleus. The kinetic energy of the α-particle will get less as its potential energy increases. At the point where all its kinetic energy has been converted into potential energy, the particle will be momentarily stationary, before the Coulomb repulsion force starts moving it away again.

The electric potential around the gold nucleus is calculated from

$$V = \frac{Q}{4\pi\varepsilon_0 r}$$

$$\therefore V_{gold} = \frac{79e}{4\pi\varepsilon_0 r}$$

Remember, the potential is the work done per unit charge in bringing a particle from infinity to a distance r from the object. So, by using Equation 3.1 for an α-particle of charge $2e$, the amount of work done E_W is

$$E_W = Q_\alpha V_{gold}$$
$$\therefore E_W = 2e \times \frac{79e}{4\pi\varepsilon_0 r} = 1.00 \times 10^{-13}$$
$$\text{So } r = \frac{158e^2}{4\pi\varepsilon_0 \times 1.00 \times 10^{-13}}$$
$$\therefore r = 3.64 \times 10^{-13} \text{ m}$$

At distance r = 3.64 × 10^{-13} m from the gold nucleus, all the kinetic energy of the α-particle has been turned into potential energy, and so this is the closest to the nucleus that the α-particle can get.

. . . .

Rutherford scattering

Suppose a Rutherford scattering experiment was carried out firing a beam of protons at a gold foil. What would be the closest that a proton could get to a nucleus if it had a non-relativistic energy of 8.35 × 10^{-14} J?

Go online

(m_p = 1.67 × 10^{-27} kg, e = 1.60 × 10^{-19} C, ε_0 = 8.85 × 10^{-12} F m^{-1})

. . . .

3.5 The electronvolt

The electronvolt (eV) is a unit of energy commonly used in high energy particle physics. The electronvolt is equal to the kinetic energy gained by an electron when it is accelerated by a potential difference of one volt. So an electron in a high energy accelerator moving through a potential difference of 4 000 000 V will gain 4 000 000 eV of energy.

The work done when a particle of charge Q moves through a potential difference V is given by $E_W = QV$. The charge on one electron is 1.6 × 10^{-19} C and so one electronvolt can be expressed in joules as follows.

$$E_W = QV$$
$$E_W = 1.6 \times 10^{-19} \times 1$$
$$E_W = 1.6 \times 10^{-19} \text{J}$$

Example

Problem:

The Large Hadron collider was designed to run at a maximum collision energy of 14 TeV. Express this in joules.

Solution:

$$E_W = QV$$
$$E_W = 1.6 \times 10^{-19} \times 14 \times 10^{-12}$$
$$E_W = 2.2 \times 10^{-6} \text{J}$$

....

3.6   Summary

In this topic we have seen that charged particles are accelerated by electric fields, gaining kinetic energy. An electric field can be used to deflect the path of a charged particle or a beam of such particles.

Summary

You should now be able to:

- describe the energy transformation that takes place when a charged particle is moving in an electric field;

- carry out calculations using $E_w = QV$;

- define an electronvolt;

- describe the motion of a charged particle in a uniform electric field, and use the kinematic relationships to calculate the trajectory of this motion;

- perform calculations to solve problems involving charged particles in electric fields, including the collision of a charged particle with a stationary nucleus.

3.7   Extended information

Web links

There are web links available online exploring the subject further.

....

3.8   Assessment

End of topic 3 test

The following test contains questions covering the work from this topic.

Go online

The following data should be used when required:

Charge on electron e	-1.60×10^{-19} C
Mass of an electron m_e	9.11×10^{-31} kg
Speed of light in vacuum c	3.00×10^{8} m s^{-1}
Mass of an α-particle	6.65×10^{-27} kg
Charge of an α-particle	$+3.20 \times 10^{-19}$ C

Q10: An electron is accelerated through a potential of 280 V.

Calculate the resultant increase in the kinetic energy of the electron.

_____ J

. . . .

Q11: An electron is accelerated from rest through a potential of 75 V.

Calculate the final velocity of the electron.

_____ m s^{-1}

. . . .

Q12: Consider a particle of charge 7.7 μC and mass 2.5×10^{-4} kg entering an electric field of strength 6.4×10^5 N C^{-1}.

Calculate the acceleration of the particle.

_____ m s^{-2}

. . . .

Q13: An electron travelling horizontally with velocity 1.55×10^6 m s^{-1} enters a uniform electric field, as shown below.

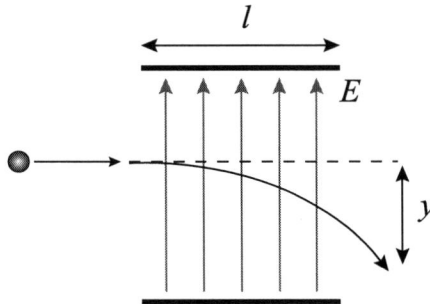

The electron travels a distance $l = 0.0200$ m in the field and the strength of the field is 160 N C^{-1}.

1. Calculate the vertical component of the electron's velocity when it emerges from the E-field.

   _____ m s^{-1}

2. Calculate the vertical displacement y of the electron.

   _____ m

. . . .

Q14: In a Rutherford scattering experiment, an α-particle (charge +2e) is fired at a stationary gold nucleus (charge +79e).

Calculate the work done by the α-particle in moving from infinity to a distance 5.45×10^{-13} m from the gold nucleus.

_____ J

. . . .

Topic 4

Magnetic fields (Unit 3)

Contents

Prerequisite knowledge

- *An understanding of the force on a charged particle placed in a magnetic field (Unit 2 - Topic 3).*

- *An understanding of the concept of electrical field (Topics 1 to 3).*

- *An understanding of the concept of gravitational field (Unit 1 - Topic 5).*

- *Basic geometrical and algebraic skills.*

Learning objectives

By the end of this topic you should be able to:

- *state that electrons are in motion around atomic nuclei and individually produce a magnetic effect;*

- *state that ferromagnetism is a magnetic effect in which magnetic domains can be made to line up, resulting in the material becoming magnetised;*

- state that iron, nickel, cobalt and some compounds of rare earth metals are ferromagnetic;

- sketch the magnetic field patterns around permanent magnets and the Earth;

- state that a magnetic field exists around a moving charge in addition to its electric field;

- sketch the magnetic field patterns around current carrying wires and current carrying coils;

- state that a charged particle moving across a magnetic field experiences a force;

- explain the interaction between magnetic fields and current in a wire;

- state the relative directions of current, magnetic field and force for a current-carrying conductor in a magnetic field;

- describe how to investigate the factors affecting the force on a current-carrying conductor in a magnetic field;

- use the relationship $F = IlB \sin \theta$ for the force on a current-carrying conductor in a magnetic field;

- define the unit of magnetic induction, the tesla (T);

- state and use the expression $B = \frac{\mu_0 I}{2\pi r}$ for the magnetic field B due to a straight current-carrying conductor;

- compare gravitational, electrostatic, magnetic and nuclear forces.

4.1   Introduction

In Unit 2 - Topic 3 Particles From Space, you met the terms magnetic field and magnetic induction. You studied the force that acts on a charged particle moving in a magnetic field and you looked at the effect the Earth's magnetic field has on cosmic rays.

In this topic we will find out why some materials are attracted to magnets and others are not. We will look more closely at the magnetic field patterns between magnetic poles, around solenoids and around the Earth. We will describe the magnetic force on a current-carrying conductor by using a field description. We will then investigate how magnetic induction varies with distance from a current carrying wire. Finally, we will compare gravitational, electrostatic, magnetic and nuclear forces.

4.2   Magnetic forces and fields

An atom consists of a nucleus surrounded by moving electrons. Since the electrons are charged and moving, they create a magnetic field in the space around them. Some atoms have magnetic fields associated with them and behave like magnets. Iron, nickel and cobalt belong to a class of materials that are **ferromagnetic**. In these materials, the magnetic fields of atoms line up in regions called **magnetic domains**. If the magnetic domains in a piece of ferromagnetic material are arranged so that most of their magnetic fields point the same way, then the material is said to be a magnet and it will have a detectable magnetic field.

Each small arrow represents the magnetic field in a magnetic domain. A refrigerator magnet is an everyday example of ferromagnetism and this property has many applications in modern technology, such as the magnetic storage in hard disks.

Figure 4.1: Magnetic domains

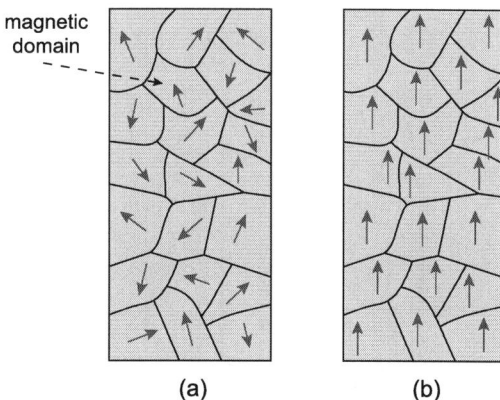

(a)   (b)

(a) Unmagnetised material where the magnetic domains cancel, (b) Magnetised material where the domains align

....

Magnetic domains

Go online

There is an online animation showing how magnetic domains respond to an outside magnetic field.

....

You may recall the magnetic field pattern around a bar magnet from earlier on in the course.

Figure 4.2: The field pattern around a bar magnet

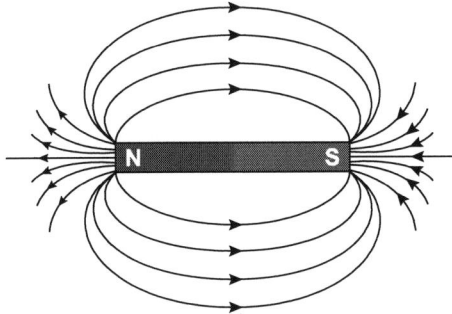

....

Remember that the convention is to draw the field direction as outward from the North pole and in to the South pole of the magnet. The distance between the lines increases as you move further from the magnet, since the magnetic field strength decreases.

The magnetic field pattern for a combination of magnets is shown below.

Figure 4.3: The field pattern around a pair of bar magnets - two opposite poles

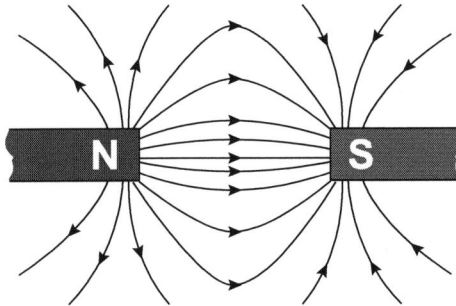

....

Figure 4.4: The field pattern around a pair of bar magnets - two like poles

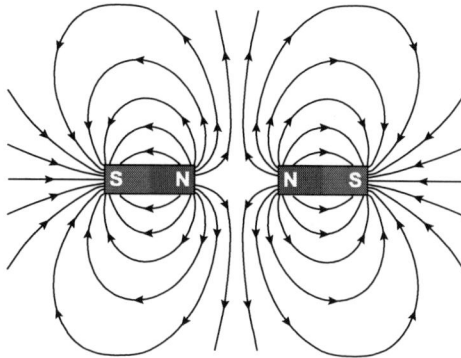

. . . .

Your teacher may provide you with equipment to confirm this to be the case.

In earlier topics, we introduced the concept of the gravitational field associated with a mass (Rotational Motion and Astrophysics - Topic 5) and the electric field associated with a charge (Electromagnetism - Topic 1).

An electric force exists between two or more charged particles whether they are moving or not.

We can explain magnetic interactions by considering that moving charges or currents create magnetic fields in the space around them, and that these magnetic fields exert forces on any other moving charges or currents present in the field.

You may recall that an interesting consequence of this is the Earth itself has a magnetic field. The flow of liquid iron within its molten core generates electric currents, which in turn produce a magnetic field. The magnetic field pattern around the Earth is similar to that of a bar magnet, but it is worth noting that the geographical north pole acts like a magnetic south pole. Therefore, the magnetic field lines actually point towards the geographical north pole.

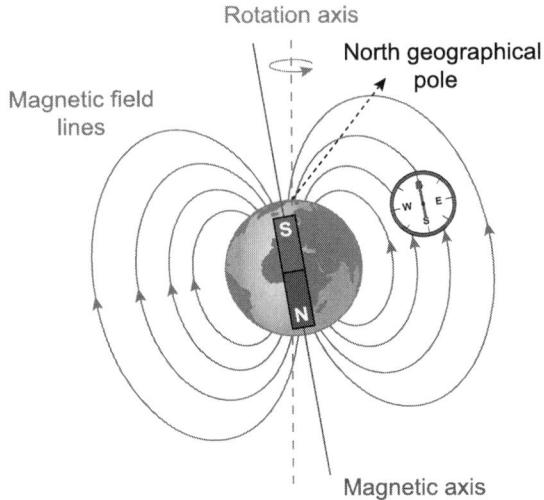

Quiz: Magnetic fields and forces

Go online

Q1:   Which one of the following statements about magnets is correct?

a)   All magnets have one pole called a monopole.

b)   All magnets are made of iron.

c)   Ferromagnetic materials cannot be made into magnets.

d)   All magnets have two poles called positive and negative.

e)   All magnets have two poles called north and south.

. . . .

Q2:   Which of the following statements about magnetic field lines is/are correct?
Magnetic field lines:
(i) are directed from the north pole to the south pole of a magnet.
(ii) only intersect at right angles.
(iii) are further apart at a weaker place in the field.

a)   (i) only

b)   (ii) only

c)   (iii) only

d)   (i) and (ii) only

e)   (i) and (iii) only

. . . .

Q3: Which of the following statements about the Earth's magnetic field is/are correct?
The Earth's magnetic field:
(i) is horizontal at all points on the Earth's surface.
(ii) has a magnetic north pole at almost the same point as the geographic north pole.
(iii) is similar to the field of a bar magnet.

a) (i) only
b) (ii) only
c) (iii) only
d) (i) and (iii) only
e) (i), (ii) and (iii)

. . . .

4.3   Magnetic field around a current-carrying conductor

Current is a movement of charges. We have just seen that there is a magnetic field
round about moving charges, so there must be a magnetic field round a wire carrying a
current. This effect was first discovered by the Danish physicist Hans Christian Oersted
(1777 - 1851). Oersted was in fact the first person to link an electric current to a magnetic
compass needle.

Oersted's experiment

*This example experiment, in common with all of this Scholar course, marks the
current as the direction of electron flow.*

Go online

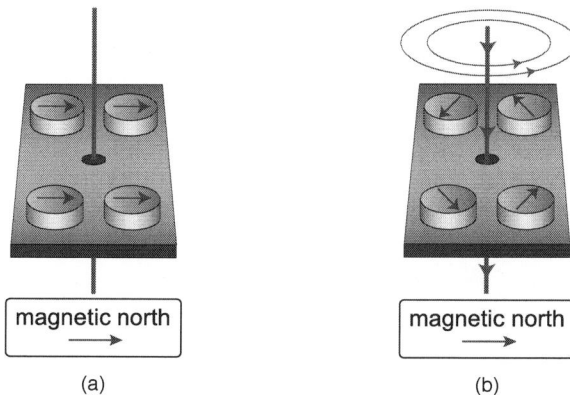

magnetic north
⟶

(a)

magnetic north
⟶

(b)

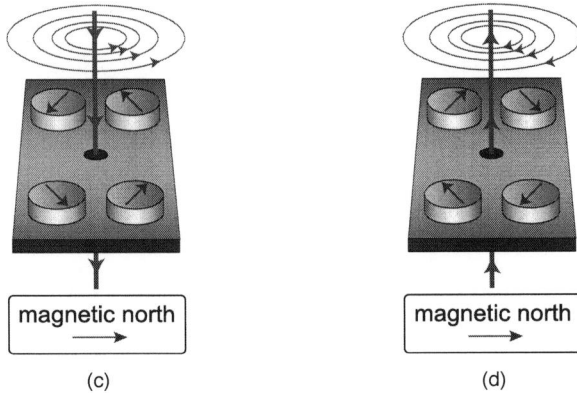

(c)   (d)

a) When there is no current, there is no magnetic field around the wire and the compass needles react to the magnetic field around the earth.

b) A current is now passed through the wire.

1. When the current is switched on what is the shape of the magnetic field?

c) The magnitude of the current is now increased.

2. When the current is increased what happens to the strength of the magnetic field?

d) The direction of the flow of current is now reversed.

3. When the current is reversed what happens to the direction of the magnetic field?

. . . .

A current through a wire produces a circular field, centred on the wire as shown in Figure 4.5. *I* shows the direction of electron current flow (current flows in the direction negative to positive).

Figure 4.5: The magnetic field around a straight wire

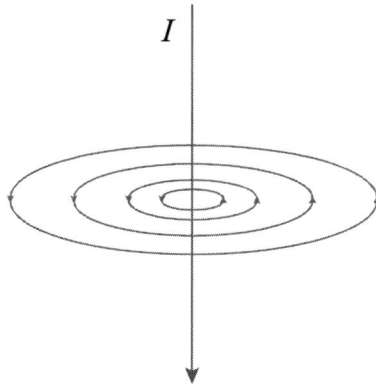

The direction of the magnetic field can be found by using the left-hand grip rule (for electron current), as follows:

Point the thumb of the left hand in the direction of the current, that is the direction in which the electrons are moving. The way the fingers curl round the wire when making a fist is the way the magnetic field is directed. This rule is sometimes known as the left-hand grip rule.
A way to remember this is thu**M**b = Motion of electrons and **F**ingers = Field lines.

Figure 4.6: The left-hand grip rule for electron current

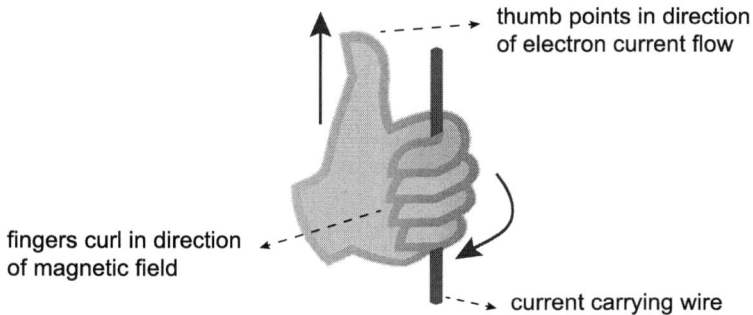

The magnetic field associated with a single straight length of wire is not very strong. If the wire is shaped into a flat circular coil, then the magnetic field inside the coil is more concentrated. The field pattern caused by a current in a flat circular coil of wire is shown in Figure 4.7.

Figure 4.7: The magnetic field pattern caused by current in a flat circular coil

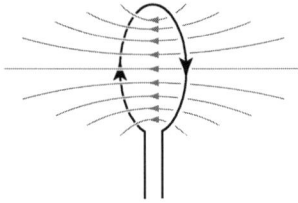

The magnetic field can be further strengthened by winding a wire into a long coil, known as a solenoid. The magnetic field pattern caused by current in a long solenoid is shown in Figure 4.8. Another version of the left-hand grip rule can be used to predict the direction of the magnetic field associated with both the flat circular coil and the long solenoid.

In this case, curl the fingers of the left hand round the coil or the solenoid in the direction of the electron current. The thumb then points towards the north end of the magnetic field produced in the solenoid. See Figure 4.7, Figure 4.8 and Figure 4.9.

Figure 4.8: The left-hand rule for solenoids

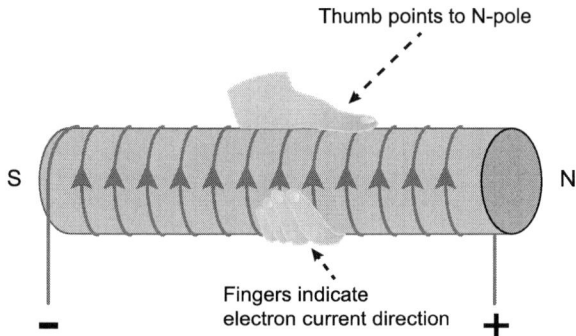

Thumb points to N-pole

S   N

− Fingers indicate electron current direction +

Figure 4.9: The magnetic field pattern caused by current in a long solenoid

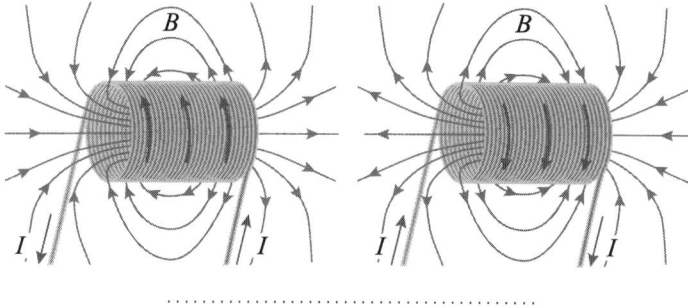

Magnetic field lines around a solenoid

There is an online animation which will help you to understand the magnetic field lines around a solenoid.

Go online

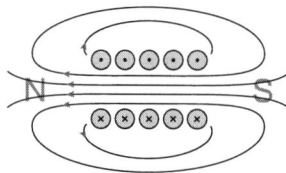

4.4   Magnetic induction

So far we have used a magnetic field description without quantifying it. We will now use one of the effects of a magnetic field to do just that. The symbol that is used for the magnetic field is B. The magnetic field is a vector quantity, and so has a direction associated with it. The direction of the field at any position is defined as the way that the north pole of a compass would point in the field at that position. There are other names that are used for magnetic field - magnetic flux density, magnetic induction or magnetic B-field. They all come about from different approaches to an understanding of magnetic fields.

The unit for **magnetic induction**, the tesla (T), is obtained from the force on a conductor in a magnetic field. One tesla is the magnetic induction of a magnetic field in which a conductor of length one metre, carrying a current of one ampere perpendicular to the field is acted on by a force of one newton.

$$1\ T = 1\ N\ A^{-1}\ m^{-1}$$

As in all areas of Physics, it is useful to have a 'feel' for the quantities that you are dealing with. The order of magnitude values shown in Table 4.1 might be of use in gaining an understanding of magnetic fields.

Table 4.1: Typical magnetic field values

Situation	Magnetic field (T)
Magnetic field of the Earth	5×10^{-5}
At the poles of a typical fridge magnet	1×10^{-3}
Between the poles of a large electromagnet	1.00
In the interior of an atom	10.0
Largest steady field produced in a laboratory	45.0
At the surface of a neutron star (estimated)	1.0×10^{8}

. . . .

4.5 Force on a current-carrying conductor in a magnetic field

The forces that a magnetic field exerts on the moving charges in a conductor are transmitted to the whole of the conductor and it experiences a force that tends to make it move. Consider a **current-carrying conductor** that is in a uniform magnetic field B, as in Figure 4.10.

Figure 4.10: A current-carrying conductor in a magnetic field

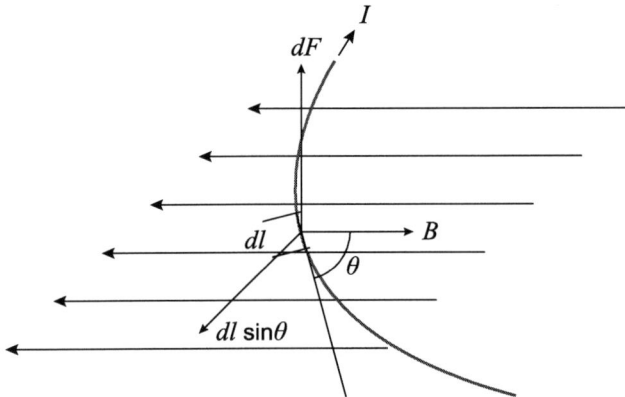

. . . .

The force dF on a small length dl of the conductor is proportional to the current I, the magnetic induction B, and the component of dl perpendicular to the magnetic field, that is $dl \sin \theta$.

$$dF = BI \, dl \sin \theta$$

(4.1)

. . . .

where θ is the angle between the length dl of the conductor and the magnetic field B.

For a straight conductor of length l in a uniform field B, the force on the conductor becomes

$$F = BIl \sin \theta$$

(4.2)

. . . .

If the conductor, and so also the current, is perpendicular to the field, then $\sin\theta = \sin 90° = 1$ and so the force is a maximum and is given by

$$F = BIl$$

If the conductor is parallel to the field, $\sin\theta = 0$ and the force is zero.

Example

Problem:

Calculate the magnitude of the force on a horizontal conductor 10 cm long, carrying a current of 20 A from south to north, when it is placed in a horizontal magnetic field of magnitude 0.75 T, directed from east to west.

Solution:

$$F = B\,I\,l\sin\theta$$
$$= 0.75 \times 20 \times 0.1 \times 1$$
$$= 1.5\ \text{N}$$

If a charge's velocity vector is not perpendicular to the magnetic field, then the component of v perpendicular to the field $v_\perp$ must be used in the equation $F = B\,I\,l\sin\theta$.

The direction of the force is at right angles to the plane containing l and B.
You may recall from **Unit 2** that the direction of this force can be established using the right hand rule.

. . . .

Figure 4.11: Direction of force on electrons

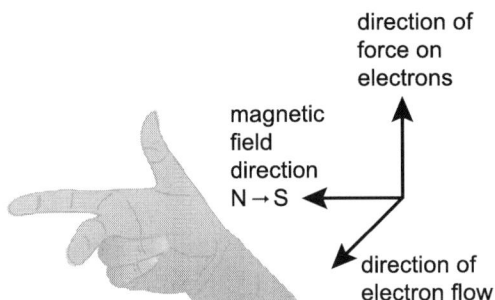

If the second finger points in the direction the electrons are flowing and the first finger points from north to south in the magnetic field then the thumb gives the direction of the force acting on the electrons.

Some people remember this right-hand rule as

- **T**humb for **t**hrust (force)
- Fore **F**inger for **F**ield, N $\rightarrow$ S
- **C**entre finger for **c**urrent

Force on a current-carrying conductor

At this stage there is an online activity which demonstrates the force exerted on a current-carrying conductor placed in the field of a horseshoe magnet.

Go online

. . . .

Force-on-a-conductor balance

The relationship between the magnetic induction B between two magnets and the force

on a current-carrying conductor can be verified using a current balance shown in Figure
4.12

Figure 4.12: Measuring the force on a current-carrying conductor

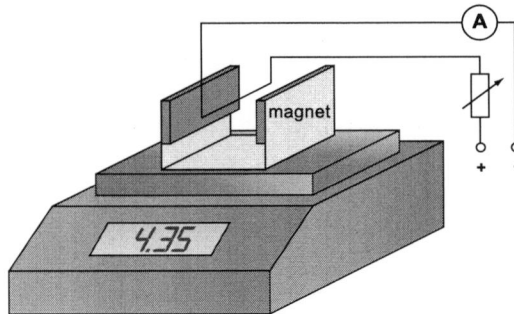

The balance is zeroed with no current in the wire. When a current is passed through the wire, the force F exerted by the wire on the magnet is seen as an apparent increase or decrease in the mass of the magnet Δm. This change in apparent mass is caused by a force of $\Delta m \; g$ newtons.

Force-on-a-conductor balance

Go online

At this stage there is an online activity which investigate the factors affecting the force on a conductor in a magnetic field.

4.5.1   The electric motor

The electric motor is device that makes use of the magnetic torque on a coil suspended in a magnetic field.

Consider the simple motor shown in Figure 4.13.

Figure 4.13: The simple electric motor

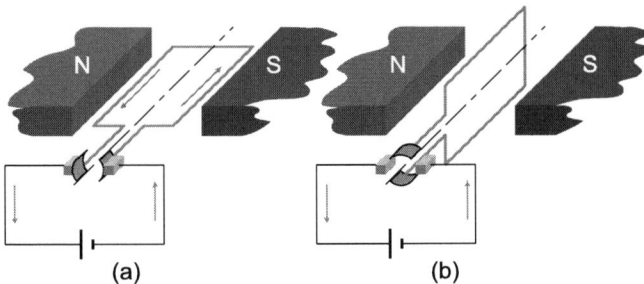

(a)   (b)

The coil, in this simple case consisting only of one turn, is called the rotor. It is free to rotate about an axis through its centre. The coil is placed in a magnetic field, which at the moment we will consider to be uniform. A current is fed into and out of the coil from an external circuit containing a source of e.m.f. through two brushes which contact with a commutator. The commutator consists of a split ring with each half connected to each end of the coil.

In Figure 4.13 (a) it can be seen that there is a force on each of the long sides of the coil. Since the current in each of these two sides is in opposite directions, these two forces supply a magnetic torque to the coil that makes it move anti-clockwise when looking in the direction shown.

This magnetic torque continues to move the coil round until it reaches the position shown in Figure 4.13 (b). At this position, if the current continued in the same direction, there would no longer be a torque on the coil (although there are still forces on each of the sides, these forces now act in opposite directions along the same line of action and so the torque has reduced to zero). Momentarily at this position, however, both sides of the commutator are in contact with both of the brushes. This stops the current in the coil. The inertia of the coil takes it slightly beyond the equilibrium position shown in Figure 4.13 (b) and this results in each brush again only connecting with one side of the commutator, restarting the current.

Although the sides of the coil have now physically changed positions, the current always enters the side of the coil that is nearest to the north pole and always leaves by the side nearest to the south pole. So the current reverses direction in the rotor every half-revolution and this current reversal, coupled with the rotation of the coil, ensures that the magnetic torque is always in the same sense.

The simple electric motor

At this stage there is an online activity.

. . . .

Go online

4.5.2 The electromagnetic pump

Consider a conducting fluid in a pipe with an electric current passing through it in a direction that is at right angles to the pipe. If the pipe is placed in a magnetic field that is at right angles to both the direction of the current and the pipe, then the fluid will experience a force along the length of the pipe, as shown in Figure 4.14. This will cause the fluid to flow along the pipe under the action of the magnetic force, with no external mechanical force applied to it. The twin benefits of this type of pumping action compared to a conventional mechanical pump are that the system is completely sealed and there are no moving parts other than the fluid itself.

Figure 4.14: The electromagnetic pump

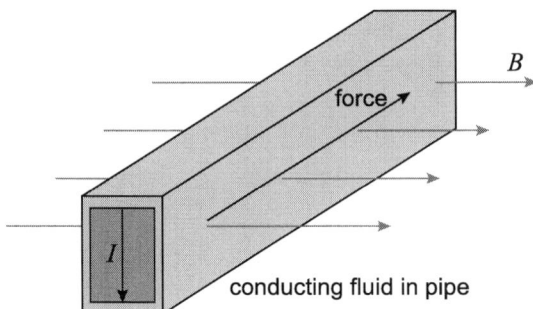

This type of pump is widely used in nuclear reactors to transport the liquid metal sodium that is used as a coolant from the reactor core to the turbine. More recently, electromagnetic pumps have been used in medical physics to transport blood in heart-lung machines and artificial kidney machines. Blood transported in this way can remain sealed and so the risk of contamination is reduced. There is also less damage to the delicate blood cells than is caused by mechanical pumps that have moving parts.

Quiz: Current-carrying conductors

Go online

Q4: The force on a conductor in a magnetic field is measured when the conductor is perpendicular to the field. Changes are made to the magnitude of the field, the current and the length of the conductor in the field.

In which one of the following situations is the force the same as the original force?

a) field halved, current the same, length the same
b) field halved, current halved, length halved
c) field doubled, current doubled, length doubled
d) field the same, current the same, length doubled
e) field the same, current doubled, length halved

. . . .

Q5: The force on a conductor in a magnetic field is measured when the conductor is perpendicular to the field.

Through what angle must the conductor be rotated in the direction of the magnetic field to reduce the force to half its original value?

a) 0°
b) 30°
c) 45°
d) 60°
e) 90°

. . . .

Q6: Which is the correct description for the magnetic field around a long straight wire carrying a current?

a) radial, directed out from the wire
b) radial, directed in to the wire
c) uniform at all points
d) circular, increasing in magnitude with distance from the wire
e) circular, decreasing in magnitude with distance from the wire

. . . .

Q7: Which of the following is equivalent to the unit of magnetic induction, the tesla?

a) $N\,A\,m^{-1}$
b) $N\,A^{-1}\,m^{-1}$
c) $N\,m^{-1}$
d) $N\,m\,A^{-1}$
e) $N\,m\,rad^{-1}$

. . . .

4.6 The relationship between magnetic induction and distance from a current-carrying conductor

Earlier in the topic, we saw that the magnetic field around a long straight wire carrying a current is circular, and is centred on the wire. A Hall probe, smartphone or search coil can be used to measure the magnitude of the field at various points. Such an investigation shows that the magnitude of the field, B, is directly proportional to the current, I, in the wire and is inversely proportional to the distance, r, from the wire.

$$B \propto \frac{I}{r}$$

The constant of proportionality in this relationship is written as $\mu_0/2\pi$, so the relationship becomes

$$B = \frac{\mu_0 I}{2\pi r}$$

(4.3)

. . . .

The constant μ_0 in Equation 4.3 is called the **permeability of free space** and it has a value of $4\pi \times 10^{-7}$ H m^{-1} (or T m A^{-1}).

μ_0 is the counterpart in magnetism to ε_0, the permittivity of free space, that appears in electrostatics. You will also have noticed that μ_0 appears in the numerator of the expression for magnetic induction, while ε_0 appears in the denominator of the expression for electric field ($E = \frac{Q}{4\pi\varepsilon_0 r^2}$).

This is partly explained by the fact that any insulating material placed in an electric field decreases the magnitude of the field, so relative permittivity appears as a divisor. On the other hand, inserting a ferromagnetic material in a magnetic field increases the magnitude of the field. Hence relative permeability appears as a multiplier.

Example

Problem:

Calculate the magnitude of the magnetic field at a point in space 12 cm from a long straight wire that is carrying a current of 9.0 A.

Solution:

We are given that the current I is 9.0 A and we want to calculate B at a point where r is 0.12 m.

$$B = \frac{\mu_0 I}{2\pi r}$$
$$\therefore B = \frac{4\pi \times 10^{-7} \times 9.0}{2\pi \times 0.12}$$
$$\therefore B = 1.5 \times 10^{-5} \text{ T}$$

. . . .

The hiker

Go online

A hiker is standing directly under a high voltage transmission line that is carrying a current of 500 A in a direction from north to south. The line is 10 m above the ground.

a) Calculate the magnitude of the magnetic field where the hiker is standing.

b) Calculate the minimum distance the hiker has to walk on horizontal ground to be able to rely on the reading given by his compass, assuming that any external magnetic field greater than 10% of the value of the Earth's magnetic field adversely affects the operation of a compass.

Take the magnitude of the Earth's magnetic field to be 0.5 x 10^{-4} T.

. . . .

4.7   Comparison of forces

In our everyday lives there are two forces of nature that shape the world around us - electromagnetic and gravitational forces. Any other forces are just different manifestations of these forces. For example, you might consider the force involved when you stretch an elastic band. The tension in the band comes about because there are electrostatic attractions between the atoms in the band. As you pull on the band, these electric forces supply an opposing force. While you may think of this as a 'mechanical' force, fundamentally this is an electromagnetic interaction.

Example

Problem:

Let's consider the interesting situation where both Coulomb and gravitational forces are present - which of the forces is dominant? For example, in a hydrogen atom we have two charged particles of known mass, so both forces are present. Is it the Coulomb force or the gravitational force that keeps them together as a hydrogen atom?

Solution:

The proton and electron which make up a hydrogen atom have equal and opposite charges, $e = 1.60 \times 10^{-19}$ C. The mass of a proton $m_p = 1.67 \times 10^{-27}$ kg and the mass of an electron $m_e = 9.11 \times 10^{-31}$ kg. The average separation between proton and electron $r = 5.29 \times 10^{-11}$ m.

Coulomb Force	Gravitational Force
$F = \dfrac{Q_1 Q_2}{4\pi\varepsilon_0 r^2}$	$F = G \dfrac{m_1 m_2}{r^2}$
$\therefore F = \dfrac{1.60 \times 10^{-19} \times \left(-1.60 \times 10^{-19}\right)}{4\pi\varepsilon_0 \times \left(5.29 \times 10^{-11}\right)^2}$	$\therefore F = G \times \dfrac{1.67 \times 10^{-27} \times 9.11 \times 10^{-31}}{\left(5.29 \times 10^{-11}\right)^2}$
$\therefore F = \dfrac{-2.56 \times 10^{-38}}{4\pi\varepsilon_0 \times 2.798 \times 10^{-21}}$	$\therefore F = 6.67 \times 10^{-11} \times \dfrac{1.521 \times 10^{-57}}{2.798 \times 10^{-21}}$
$\therefore F = -8.23 \times 10^{-8}$ N	$\therefore F = 3.63 \times 10^{-47}$ N

Remember the minus sign in front of the Coulomb force just indicates that we have an attractive Coulomb force (oppositely charged particles). The gravitational force is also attractive, but by convention the negative sign in omitted. Our calculations show that for atomic hydrogen, the Coulomb force is greater than the gravitational force by a factor of 10^{39}! So in this case the gravitational force is negligible compared to the electrostatic (Coulomb) force.

. . . .

Now let us consider what happens inside the nucleus of an atom.

Although we can describe everyday phenomena in terms of electromagnetic or gravitational forces, you will recall from the Higher course that we need to consider other forces when describing nuclei. To understand this, let us consider a helium nucleus, consisting of two positively charged protons and two uncharged neutrons. Clearly there is an electrostatic repulsion between the protons. There is also a gravitational attraction between them. We can carry out an order-of-magnitude calculation to compare these forces.

Go online

Electrostatic and gravitational forces

Two protons in a helium nucleus are separated by around 10^{-15} m. Given the following data, perform an order-of-magnitude calculation to determine the ratio F_C/F_G of the Coulomb (electrostatic) and gravitational forces that act between the two protons. You will need to know the values of the permittivity of free space and the gravitational constant:

charge on a proton	$+1.6 \times 10^{-19}$ C
mass of a proton	1.67×10^{-27} kg
permittivity of free space ε_0	8.85×10^{-12} F m^{-1}
gravitational constant G	6.67×10^{-11} N m^2 kg^{-2}

. . . .

These order-of-magnitude calculations shows that the electrostatic force is very much greater than the gravitational force. If this is the case, why doesn't this repulsive force cause the nucleus to split apart? The reason is that there is another force, called the **strong nuclear force**, that acts between any two nucleons (protons or neutrons) in a nucleus. This force, usually just called the strong force, can act between two protons, two neutrons, or a proton and a neutron, and has almost the same magnitude in each case.

How does the strong force compare to the electromagnetic and gravitational forces? The main difference is that the strong force is a short-range force. In fact its range is $< 10^{-14}$ m. This means that unless a particle is closer than about 10^{-14} m to another particle, the strong force between them is effectively zero. This is in contrast to the other two forces, both of which are long-range. For example, the gravitational force is the main interaction between planets and stars, and is effective over many millions of kilometres. The electrical force between charged objects follows a similar $1/r^2$ dependence.

In terms of strength, over a short range, the strong force is very much greater than the gravitational force, which is negligibly small between particles of such low mass. The strong force overcomes the electrostatic repulsion between protons, and prevents the nucleus from disintegrating.

Another point to note is that electrons are not affected by the strong force. A proton and an electron, or a neutron and an electron, would not interact via the strong force.

You may also recall from the Higher course that the weak nuclear force is responsible for beta decay.

One of the biggest challenges in theoretical physics is to explain fully all four fundamental forces. Table 4.2 compares these four forces.

Table 4.2: Comparison of the forces of nature

Force	Relative magnitude	Range (m)	Example
strong nuclear	1	$< 10^{-14}$	nucleons in a nucleus
electromagnetic	$\sim 10^{-2}$	∞	majority of everyday 'contact' forces
weak nuclear	$\sim 10^{-5}$	$\sim 10^{-18}$	β-decay of a nucleus
gravitational	$\sim 10^{-38}$	∞	very large masses, e.g. planets

. . . .

Gravity is clearly a much weaker force than the electric force. Furthermore, it is always attractive, whereas the electric force can be either attractive or repulsive. Despite these obvious differences, during the 1700s scientists noticed a great deal of similarity between the two forces, leading to speculation that they were perhaps really just manifestations of the same thing. Table 4.3 highlights the similarities between gravitational and electric forces. We shall further explore the unification of forces in **Topic 7**.

Table 4.3: Comparison of gravitational and electric forces

Concept	Gravitational field	Electric field
Force	Force between point masses obeys an inverse square law $$F = \frac{GMm}{r^2}$$ Only attraction	Force between point charges obeys an inverse square law $$F = \frac{Q_1 Q_2}{4\pi\varepsilon_0 r^2}$$ Repulsion or attraction
Field lines	A radial field surrounds a point mass and the field lines are drawn towards the mass.	A radial field surrounds a point charge and the field lines are drawn in the direction a positive charge would move.

Concept	Gravitational field	Electric field
Field strength	Force per unit mass $g = F/m$ Unit is Nkg^{-1}	Force per unit charge $E = F/Q$ Unit is NC^{-1}
Field strength for point mass or charge	Field strength for point mass obeys inverse square law $g = \dfrac{GM}{r^2}$	Field strength for point charge obeys inverse square law $E = \dfrac{Q}{4\pi\varepsilon_0 r^2}$
Potential	$V = \dfrac{GM}{r}$ Joules per kg The zero of potential is at infinity from the planet.	$V = \dfrac{Q}{4\pi\varepsilon_0 r}$ Joules per Coulomb The zero of potential is at infinity from a charged object.
Potential Energy	$E_p = mV$ $E_p = \dfrac{-GMm}{r}$	$E_p = QV$ $E_p = \dfrac{Q_1 Q_2}{4\pi\varepsilon_0 r}$
Effect	Gravitational fields hold the Universe together. The gravitational force is always attractive.	Electric fields hold atoms and molecules together. The electric force may be attractive or repulsive.

. . . .

4.7.1   Millikan's oil drop experiment

The American scientist R A Millikan conducted a experiment which involved balancing the electrostatic force on an oil drop with the gravitational force acting upon it. He designed the experiment to accurately measure the charge of an electron.

Theory

The experiment works by putting a negative electric charge on a microscopic drop of oil. The motion of the oil drop is observed as it falls between two charged horizontal plates. The magnitude of the electric field between the plates can be varied, so that the drop can be held stationary, or allowed to fall with constant velocity. Knowing the electric field strength, the charge on the oil drop can be deduced. Millikan found that every drop had a charge that was equal to an integer multiple of e, and no charged drop had a charge less than e. From these observations, he concluded firstly that charge was quantised, since the drops could only have a charge that was an integer multiple of e. Secondly, he stated that e was the fundamental unit of charge - a drop which had acquired one electron had charge e, one with two extra electrons had charge $2e$ and so on.

Experiment

The experimental arrangement is shown in Figure 4.15. The oil drops are charged either by friction as they are sprayed from a fine nozzle, or by irradiating the air near the nozzle using X-rays or a radioactive source. The drops then enter the electric field via the small aperture in the upper plate.

Figure 4.15: Experimental arrangement of Millikan's oil drop experiment

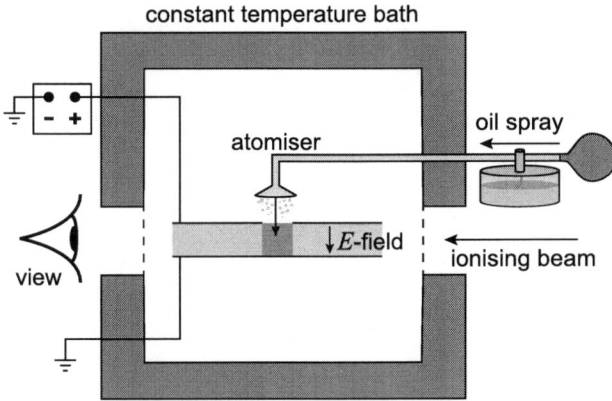

A charged drop can be held stationary in the electric field by adjusting the potential difference across the plates, or it can be allowed to fall if the pd is decreased.

Figure 4.16: Charged oil drops between charged plates

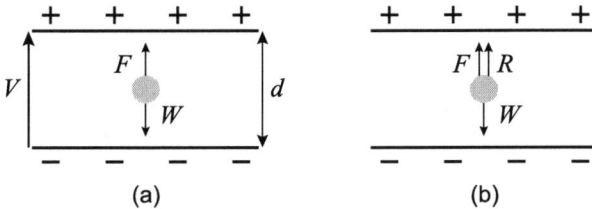

In Figure 4.16 (a) the Coulomb force F is equal to the gravitational force W acting on the drop. If the mass or radius of the drop is known, then the charge on the drop can be easily calculated. However, measuring the mass or radius of the drop is extremely difficult. Millikan devised a different method shown in Figure 4.16 (b). The drop is allowed to fall, so an additional force, air resistance R, acts on the drop. This force depends on the size of the drop and its velocity. By adjusting the pd, the drop can be made to fall at a constant velocity, so once again it is in equilibrium (Newton's First Law). If the density of the oil is known, an expression can be deduced relating the three forces acting on the drop which does not involve the mass or radius of the drop. Using this expression, the charge on the drop can be calculated.

4.8 Summary

In this topic we have seen that magnetic forces exist between moving charges. This is in addition to the electric forces that always exist between charges, moving or not.

Since a current in a wire is a movement of charges, the magnetic field around a wire was next studied. The unit for magnetic induction, the tesla, was defined. A simple way of deciding the direction of the force on a current -carrying conductor in a magnetic field was studied.

The expression for the force on a current-carrying conductor in a magnetic field was developed. It was found that the force on a conductor carrying a current in a magnetic field is proportional to both the current and the magnetic field.

We then went on to study the expression for the magnetic induction at a distance r from a current-carrying conductor. Finally, we compared the gravitational, electrostatic, magnetic and nuclear forces.

Summary

You should now be able to:

- state that electrons are in motion around atomic nuclei and individually produce a magnetic effect;

- state that ferromagnetism is a magnetic effect in which magnetic domains can be made to line up, resulting in the material becoming magnetised;

- state that iron, nickel, cobalt and some compounds of rare earth metals are ferromagnetic;

- sketch the magnetic field patterns around permanent magnets and the Earth;

- state that a magnetic field exists around a moving charge in addition to its electric field;

- sketch the magnetic field patterns around current carrying wires and current carrying coils;

- state that a charged particle moving across a magnetic field experiences a force;

Summary continued

- explain the interaction between magnetic fields and current in a wire;

- state the relative directions of current, magnetic field and force for a current-carrying conductor in a magnetic field;

- describe how to investigate the factors affecting the force on a current-carrying conductor in a magnetic field;

- use the relationship $F = IlB\sin\theta$ for the force on a current-carrying conductor in a magnetic field;

- define the unit of magnetic induction, the tesla (T);

- state and use the expression $B = \frac{\mu_0 I}{2\pi r}$ for the magnetic field B due to a straight current-carrying conductor;

- compare gravitational, electrostatic, magnetic and nuclear forces.

4.9 Extended information

Web links

There are web links available online exploring the subject further.

. . . .

4.10 Assessment

End of topic 4 test

The following test contains questions covering the work from this topic.

Go online

Q8: A long straight wire is held perpendicular to the poles of a magnet of field strength 0.311 T. Assume that the field is uniform and extends for a distance of 2.03 cm.

Calculate the force on the wire when the current in it is 4.02 A.

_____ N

. . . .

Q9: An electric power line carries a current of 1700 A.

Calculate the force on a 3.14 km length of this line at a position where the Earth's magnetic field has a magnitude of 5.35×10^{-5} T and makes an angle of 75.0° to the line.

_____ N

. . . .

Q10: A short length of wire has a mass of 13.6 grams. It is resting on two conductors that are 3.67 cm apart, at right angles to them.

The conductors are connected to a power supply which maintains a constant current of 6.83 A in the wire.

The wire is held between the poles of a horseshoe magnet that has a uniform magnetic field of 0.237 T. The wire is perpendicular to the magnetic field.

Ignoring friction, air resistance and electrical resistance, calculate the initial acceleration of the wire.

_____ m s^{-2}

....

Q11: The apparatus shown is used to investigate the magnetic induction B of an electromagnet.

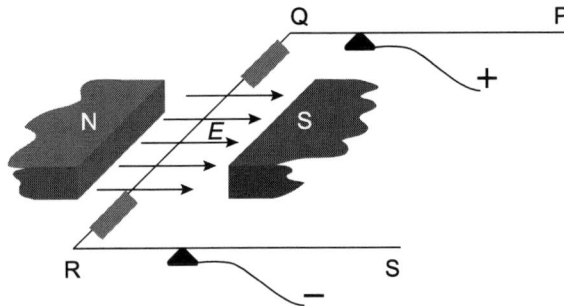

The straight wire QR has a current of 4.9 A supplied to it through the two knife edge points.

With the electromagnet switched off, the wire PQRS is balanced in a horizontal plane by hanging small masses as shown.

When the electromagnet is switched on, the mass hanging on QR has to be altered by 4.2 grams to restore PQRS to the horizontal.

The perpendicular length of QR which is in the magnetic field is 40 mm.

1. To restore PQRS to the horizontal, masses have to be
   a) removed from QR.
   b) added to QR.
2. Calculate the magnitude of the magnetic induction B.
   _____ T

....

Q12: A long straight conductor carries a steady current and produces a magnetic field of 4.1×10^{-5} T, at a distance of 11 mm.

Calculate the magnitude of the current.

_____ A

....

Q13: A lightning bolt can be considered as a straight current-carrying conductor.

Calculate the magnetic induction 18.7 m away from a lightning bolt in which a charge of 18.2 C is transferred in a time of 4.83 ms.

_____ T

. . . .

Q14: A device called a Hall probe can be used to find the current in a pipe carrying molten metal.

When the Hall probe is held perpendicular 0.53 m from the centre of the pipe, the maximum reading recorded by the probe is 1.7 mV.

The Hall probe has a sensitivity of 1000 mV T^{-1}.

1. Calculate the magnetic induction at this position.

    _____ T

2. Calculate the current in the pipe.

    _____ A

. . . .

Topic 5

Capacitors (Unit 3)

Contents

Prerequisite knowledge

- *Ohm's Law and circuit rules (CfE Higher).*

- *Capacitors (CfE Higher).*

- *r.m.s current (CfE Higher).*

- *Electric fields - Topic 1.*

Learning objectives

By the end of this topic you should be able to:

- *sketch graphs of voltage and current against time for charging and discharging capacitors in series CR circuits;*

- *define the time constant of a circuit;*

- *carry out calculations relating the time constant of a circuit to the resistance and capacitance;*

- *use graphical data to determine the time constant of a circuit;*

- *define capacitive reactance;*

- *describe the response of a capacitive circuit to an a.c. signal;*

- *use the appropriate relationship to solve problems involving capacitive reactance, voltage and current;*

- *use the appropriate relationship to solve problems involving a.c. frequency, capacitance and capacitive reactance.*

5.1 Introduction

You studied capacitors as part of the Higher course. In this topic we will now look at their behaviour in more detail, paying particular attention to the time they take to charge and discharge. To understand capacitors fully you will also need to recall the section on electric fields.

You may recall that a capacitor is a device for storing electrical energy. A capacitor consists of two conducting plates. When one plate is negatively charged and the other is positively charged, then electrical energy is stored on the capacitor. We will be reviewing how this process works, and how much energy can be stored on a capacitor.

Capacitors are important components in many electrical circuits. We will study how capacitive circuits respond to d.c. and a.c. signals. This will help us to understand some of the practical applications of capacitors.

5.2 Revision from Higher

This section will allow you to revise the content covered at CfE Higher.

5.2.1 Relationship between Q and V

A capacitor is made of two pieces of metal separated by an insulator.

When the capacitor becomes charged there will be a potential difference across the two pieces of metal.

The circuit symbol for a capacitor is

The capacitance of a capacitor is measured in farads (F) or more commonly microfarads (μF, x10^{-6}) or nanofarads (nF, x 10^{-9}). The capacitance of a capacitor depends on its construction not the charge on it or the potential difference across it.

You may recall that the charge (Q) stored by a capacitor is directly proportional to the potential difference (V) across its plates and that the constant of proportionality is the capacitance of the capacitor. The following activity will refresh your memory.

Investigating V_c and Q_c

A circuit is set up to investigate the relationship between the voltage and the charge stored on the capacitor.

Go online

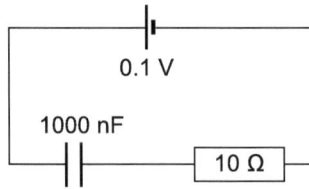

The voltage of the supply is increased (between 0.1 V and 1.0V) and the charge on the capacitor is noted.

The results obtained are used to produce the following graph.

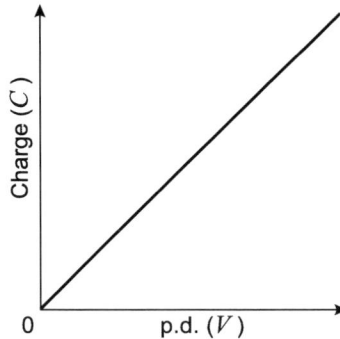

So the capacitance of a capacitor is defined by the equation

$$C = \frac{Q}{V}$$

This means that the capacitance of a capacitor is numerically equal to the amount of charge it stores when the p.d. across it is 1 volt. The unit of capacitance is the farad F, where $1F = 1 C\ V^{-1}$. It is usually more common to express the capacitance in microfarads ($1\mu F = 1\times10^{-6}$ F), nanofarads ($1nF = 1\times10^{-9}$ F) or picofarads ($1pF = 1\times10^{-12}$ F).

Example

Problem:

A 20 mF capacitor has a potential difference of 9.0 V across it. How much charge does it store?

Solution:

$$C = \frac{Q}{V}$$
$$20 \times 10^{-3} = \frac{Q}{9.0}$$
$$Q = 0.18 \ C$$

. . . .

5.2.2 Energy stored by a capacitor

Let us consider the charged parallel-plate capacitor shown in Figure 5.1.

Figure 5.1: Electric field between two charged plates

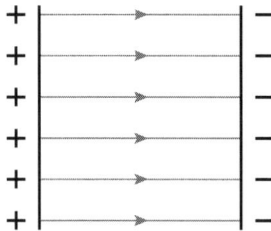

. . . .

Suppose we take an electron from the left-hand plate and transfer it to the right-hand plate. We have to do work in moving the electron since the electrical force acting on it opposes this motion. The more charge that is stored on the plates, the more difficult it will be to move the electron since the electric field between the plates will be larger.

The work that is done in placing charge on the plates of a capacitor is stored as potential energy in the charged capacitor. The more charge that is stored on the capacitor, the greater the stored potential energy.

You may recall that it can be shown that the energy stored by a capacitor is given by the following equations:

$$E = \frac{1}{2}QV \qquad\qquad E = \frac{1}{2}CV^2 \qquad\qquad E = \frac{1}{2}\frac{Q^2}{C}$$

Example

Problem:

A 750 nF capacitor is charged to 50 V. Calculate the charge and the energy it stores.

Solution:

$$E = \frac{1}{2}CV^2$$
$$E = \frac{1}{2}750 \times 10^{-9} \times 50^2$$
$$E = 9.4 \times 10^{-4} \text{ J}$$

. . . .

5.2.3 Charging and discharging capacitors in d.c. circuits

Let us now recap the behaviour of capacitors when they are connected as components in d.c. circuits. A capacitor is effectively a break in the circuit, and charge cannot flow across it. We will see now review how this influences the current in capacitive circuits.

Charging a capacitor

Go online

There is an activity online at this stage. The activity provides a circuit with a capacitor and resistor which can be altered. The shape of the output graphs is also given.

. . . .

Figure 5.2 shows a simple d.c. circuit in which a capacitor is connected in series to a battery and resistor. This is often called a series *CR* circuit.

Figure 5.2: Simple d.c. capacitive circuit

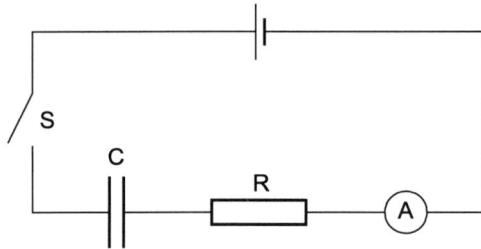

When the switch S is closed, charge can flow on to (but not across) the capacitor C. At the instant the switch is closed the capacitor is uncharged, and it requires little work to add charges to the capacitor. As we have already discussed, though, once the capacitor has some charge stored on it, it takes more work to add further charges. Figure 5.3 shows graphs of current *I* through the capacitor (measured on the ammeter) and charge *Q* on the capacitor, against time.

Increasing the R or C value increases the rise time however the final p.d. across the capacitor will remain the same. The final p.d. across the capacitor will equal the e.m.f., E, of the supply.

Figure 5.3: Plots of current and charge against time for a charging capacitor

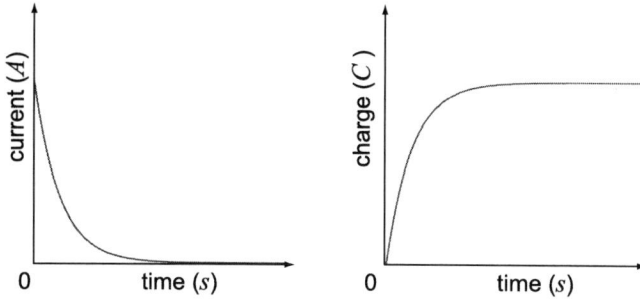

....

Since the potential difference across a capacitor is proportional to the charge on it, then a plot of p.d. against time will have the same shape as the plot of charge against time shown in Figure 5.3.

Suppose the battery in Figure 5.2 has e.m.f E and negligible internal resistance. The sum of the p.d.s across C and R must be equal to E at all times. That is to say,

$$V_C + V_R = E$$

where V_C is the p.d. across the capacitor and V_R is the p.d. across the resistor. At the instant switch S is closed there is no charge stored on the capacitor, so V_C is zero, hence $V_R = E$. The current in the circuit at the instant the switch is closed is given by

$$I = \frac{E}{R}$$

(5.1)

....

As charge builds up on the capacitor, so V_C increases and V_R decreases. This is shown in Figure 5.4.

Figure 5.4: Plots of p.d. against time for a capacitive circuit

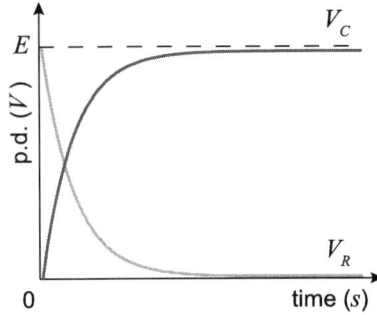

The charge and potential difference across the capacitor follow an exponential rise. (The current follows an exponential decay). The rise time (the time taken for the capacitor to become fully charged) depends on the values of the capacitance C and the resistance R. The rise time increases if either C or R increases. So, for example, replacing the resistor R in the circuit in Figure 5.2 by a resistor with a greater resistance will result in the p.d. across the capacitor C rising more slowly, and the current in the circuit dropping more slowly. We will look at this effect more closely in the next section.

Example

Problem:

Consider the circuit in Figure 5.5, in which a 40 kΩ resistor and an uncharged 220 μF capacitor are connected in series to a 12 V battery of negligible internal resistance.

Figure 5.5: Capacitor and resistor in series

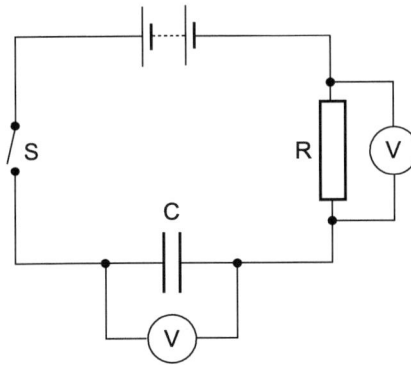

1. What is the potential difference across the capacitor at the instant the switch is closed?

2. After a certain time, the charge on the capacitor is 600 μC. Calculate the potential differences across the capacitor and the resistor at this time.

Solution:

Answer:

1. At the instant the switch is closed, the charge on the capacitor is zero, so the p.d. across it is also zero.

2. We can calculate the p.d. across the capacitor using $Q = CV$ or $C = \frac{Q}{V}$:

$$V_c = \frac{Q}{C}$$
$$\therefore V_c = \frac{600 \times 10^{-6}}{220 \times 10^{-6}}$$
$$\therefore V_c = 2.7 \text{ V}$$

Since the p.d. across the capacitor is 2.7 V, the p.d. across the resistor is 12 - 2.7 = 9.3 V.

....

5.2.4 Discharging a capacitor

Discharging a capacitor

There is an activity online at this stage showing how the capacitor charges and discharges.

Go online

....

The circuit shown in Figure 5.6 can be used to investigate the charging and discharging of a capacitor.

Figure 5.6: Circuit used for charging and discharging a capacitor

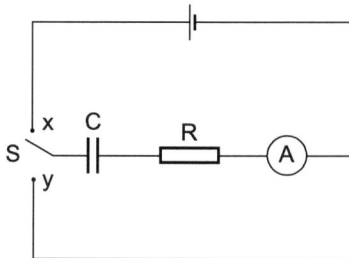

....

When the switch S is connected to x, the capacitor C is connected to the battery and resistor R, and will charge in the manner shown in Figure 5.3. When S is connected to y, the capacitor is disconnected from the battery, and forms a circuit with the resistor R. Charge will flow from C through R until C is uncharged. A plot of the current against time is given in Figure 5.7.

Figure 5.7: Current as the capacitor is charged, and then discharged

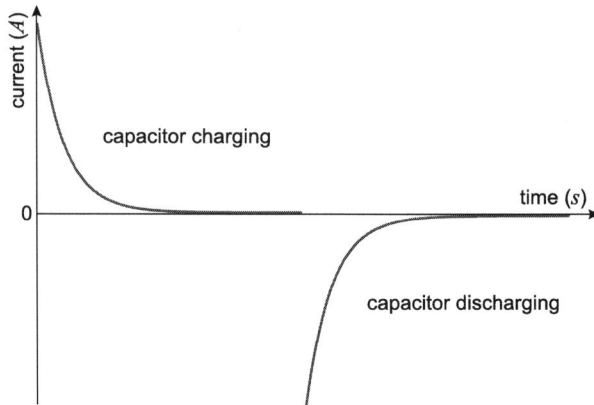

Remember that the capacitor acts as a break in the circuit. Charge is *not* flowing across the gap between the plates, it is flowing from one plate through the resistor to the other plate. Note that the direction of the current reverses when we change from charging to discharging the capacitor. The energy which has been stored on the capacitor is dissipated in the resistor.

Charging current: The initial charging current is very large. Its value can be calculated by $I = \frac{V_{\text{supply}}}{R}$.

The current is only at this value for an instant of time. As the capacitor charges, the p.d. across the capacitor increases so the p.d. across the resistor decreases causing the current to decrease.

Discharging current: The initial discharging current is very large. Its value can be calculated by $I = \frac{V_{capacitator}}{R}$

During discharge the circuit is not connect to the supply so it is the p.d. across the capacitor, not the p.d. across the supply, which drives the current. If however the capacitor had been fully charged, the initial p.d. across the capacitor would equal the p.d. across the supply.

The current is only at this value for an instant of time. As the capacitor discharges, the p.d. across the capacitor decreases so the p.d. across the resistor also decreases causing the current to decrease. When the capacitor is fully discharged, the p.d. across it will be zero hence the current will also be zero.

Figure 5.7 shows us that at the instant when the capacitor is allowed to discharge, the size of the current is extremely large, but dies away very quickly. This leads us to one of the applications of capacitors, which is to provide a large current for a short amount of time. One example is the use of a capacitor in a camera flash unit. The capacitor is charged by the camera's batteries. At the instant the shutter is pressed, the capacitor is allowed to discharge through the flashbulb, producing a short, bright burst of light.

Using the energy stored on a capacitor

At this stage there is an online activity. If however you do not have access to the internet you should ensure that you understand the following explanation.

Go online

When a lamp is lit from a d.c. supply directly it gives a steady dim energy output.

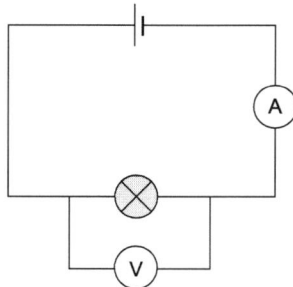

It is now connected to a capacitor and charged as shown

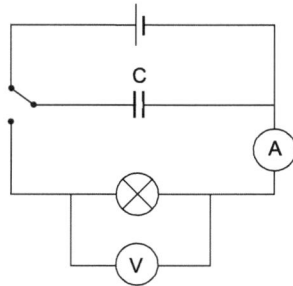

The capacitor is then discharged through the lamp by changing the switch position as shown.

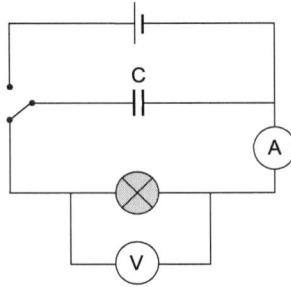

- When the capacitor powers the lamp, a large current flows for a very short period of time. This produces a bright flash of light.

- The current flows for only a short time while the capacitor discharges.

- Before the flash can be used again the capacitor must be recharged from the supply.

. . . .

The table below shows how voltage and current change during charging and discharging.

	At any instant	At start	At finish
Charging	$V_S = V_C + V_R$	$V_C = 0$	$V_C = V_S$
	$I = V_R/R$	$V_R = V_S$	$V_R = 0$
		$I_0 = V_S/R$	$I = 0$
Discharging	$V_C = V_R$	$V_C = V_S$	$V_C = 0$
	$I = V_R/R$	$V_R = V_S$	$V_R = 0$
		$I_0 = V_S/R$	$I = 0$

5.3 The time constant for a CR circuit

Now that we have gone back through the main points covered at Higher, let us now look more closely at capacitor discharge.

The graphs in the last section showed that when a capacitor is discharging, the potential difference across its plates decreases exponentially with time. As Q=CV, this means that the charge the capacitor stores must also decrease exponentially with time.

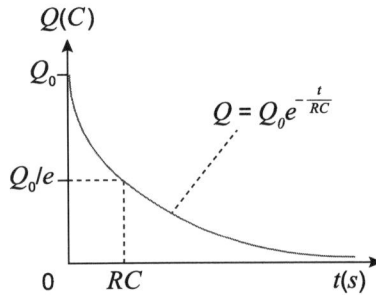

Infact, the charge left on the plates of a capacitor as it discharges is given by the equation

$$Q = Q_0 e^{-\frac{t}{RC}}$$

where Q_0 is the charge stored by the capacitor when it is fully charged, R is the resistance of the resistor and C is the capacitance of the capacitor.

If you want to see the mathematical proof for this equation, see the **For Interest Only** heading at the end of this section.

If $t = RC$ is put into the equation above, then $Q = Q_0 e^{-1}$.

So when $t = RC$, $\frac{Q}{Q_0} = \frac{1}{e}$, where $\frac{1}{e} \cong 0.37$

The quantity CR is called the **time constant** (τ) of the circuit. The time constant (τ) is the time taken for the charge on a discharging capacitor decrease to $\frac{1}{e}$ of its initial value, Q_0. In other words, the time constant (τ) is the time taken to discharge the capacitor to 37% of initial charge. The unit of time constant (τ) is the second.

After $t = 2\tau$, the value falls to $\frac{1}{e^2}$ of its initial value.

After $t = 3\tau$, the value falls to $\frac{1}{e^3}$ of its initial value.

The potential difference across the capacitor and the current in the circuit also decrease exponentially in the same manner, according to $V = V_0 e^{-\frac{t}{RC}}$ and $I = I_0 e^{-\frac{t}{RC}}$. Therefore, the time constant can also be expressed as the time taken for the current in the circuit or potential difference across the capacitor to fall to 37% of their initial values.

So, provided the resistance of the resistor and the capacitance of the capacitor are not altered, it always takes the same length of time for the Q, I and $V_{Capacitor}$ to decrease to 37% of the original value, no matter how much charge the capacitor starts with. The value of the time constant ($\tau = CR$) can be increased by increasing the value of C or R or both C and R. If CR is large, the discharge will be slow and if it is small the discharge will be fast.

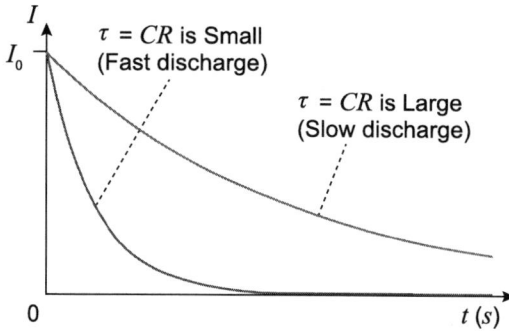

Similarly, when charging a capacitor, the charge stored by the capacitor after time t is given by the relationship $Q = Q_{max}(1 - e^{-\frac{t}{RC}})$. So the time constant is the time taken to increase the charge stored by 63% of the difference between initial charge and maximum (full) charge. The larger the resistance in series with the capacitor and the larger the capacitance of the capacitor, the longer it takes to charge.

Note that the time constant is also equal to the time taken to increase the voltage across the capacitor's plates to 63% of the maximum value.

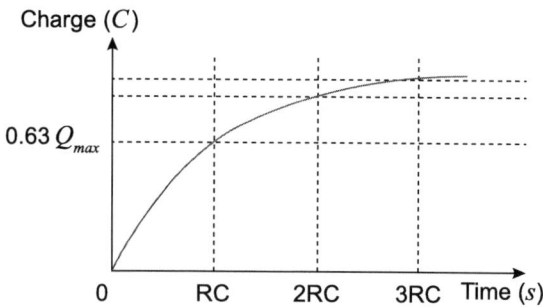

Example

Problem:

A 500 μF capacitor is fully charged from a 12 V battery and is then discharged through a 3 kΩ resistor. Calculate the time taken for the charge stored by the capacitor to decrease to 37% of the initial value.

Solution:

$$\tau = RC$$
$$\tau = 3000 \times 500 \times 10^{-6}$$
$$\tau = 1.5 \text{ s}$$

....

Example

Problem:

A 2.2 μF capacitor is connected in series to a resistor and a 6.0 V battery. It takes 0.055 s for the p.d. across its plates to increase to 3.78 V. Calculate the resistance of the resistor.

Solution:

$$\frac{3.78}{6.0} \times 100 = 63\%$$

The capacitor has been charged to 63% of the supply voltage. Therefore, the time passed must equal the time constant.

$$\tau = RC$$
$$0.055 = R \times 2.2 \times 10^{-6}$$
$$R = 25000 \ \Omega$$

. . . .

Circuits in which a capacitor discharges through a resistor are often used in electronic timers. For instance, a pelican crossing uses a capacitor to activate a sequence of traffic lights for a predetermined time when a pedestrian presses a switch. The capacitor is initially charged and is then allowed to gradually discharge. Eventually the p.d. across the capacitor will fall below a set value, triggering a switching circuit.

For Interest Only
The mathematical proof for the equation $Q = Q_0 e^{-\frac{t}{RC}}$ is not examinable. It is included here for interest only.
The current at time t during capacitor discharge is given by $I = -\frac{dQ}{dt}$. The negative sign is present because the charge stored by the capacitor decreases with time.
Substituting $V = \frac{Q}{C}$ into the equation $I = \frac{V}{R}$ and gives $I = \frac{Q}{CR}$.
Therefore, we have

$$-\frac{dQ}{dt} = \frac{Q}{CR}$$
$$\int_{Q_0}^{Q} \frac{dQ}{Q} = -\frac{1}{CR} \int_{0}^{t} dt$$
$$[ln \, Q]_{Q_0}^{Q} = -\frac{1}{CR} [t]_{0}^{t}$$
$$ln \frac{Q}{Q_0} = -\frac{t}{CR}$$
$$Q = Q_0 e^{-\frac{t}{RC}}$$

5.4 Capacitors in a.c. circuits

Capacitors oppose the flow of alternating current, as do components called inductors, which we will examine in the next topic. The opposition which a capacitor offers to a.c. current flow is called its **capacitive reactance**, X_C, and is defined by

$$X_C = \frac{V}{I}$$

Capacitive reactance is measured in Ω, as is a resistor's resistance.

We are about to explore the behaviour of a capacitor in an a.c. circuit, paying particular attention to the relationship between the a.c. frequency and the capacitive reactance. However, let us first of all find out how a resistor would behave in such a circuit.

Resistors in a.c. circuits

There is an online activity which allows you to observe the effect on the alternating current through a resistor as the frequency of the supply is altered.

Go online

. . . .

You have just observed that the ratio $\frac{V_{r.m.s.}}{I_{r.m.s.}}$ remains constant for a resistor no matter what the frequency of the a.c. supply is. This means the alternating r.m.s. current in a resistor does not change with frequency. Now let's investigate the response of a capacitor to an a.c. supply.

The capacitor and a.c.

A circuit is set up to investigate the relationship between frequency of an a.c. supply and current in a capacitive circuit.

Go online

The r.m.s. voltage of the supply is kept constant.

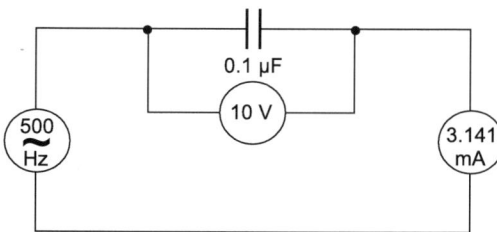

The frequency of the supply is increased and the r.m.s. current is measured.

The results obtained are used to produce the following graph.

current (mA)

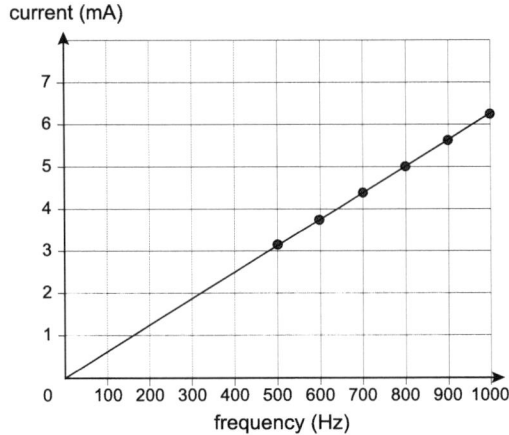

frequency (Hz)

....

As you can see the r.m.s current in a capacitive circuit is directly proportional to the frequency of the supply. This means the capacitive reactance is inversely proportional to the frequency.

You may recall that when a d.c. supply is used in a CR circuit, the current rapidly drops to zero once the switch is closed. We have just observed that in an a.c. circuit there is a steady current through the capacitor. This is because the capacitor is charging and discharging every time a.c changes direction. A CR circuit passes high frequency a.c. much better than it does low frequency a.c. or d.c. But why is this?

You should remember that charge does not flow across the plates of a capacitor. It accumulates on the plates, and the more charge that has accumulated, the more work is required to add extra charges. At all times, the *total* charge on the plates of the capacitor is zero. Charges are merely transferred from one plate to the other via the external circuit when the capacitor is charged. At low frequency, as the applied voltage oscillates, there is plenty of time for lots of charge to accumulate on the plates, which means the current drops more at low frequency (see Figure 5.3). At high frequency, there is only a short time for charge to accumulate on the plates before the direction of the current is reversed, and the capacitor discharges.

Applications

The fact a capacitor passes high frequency a.c. much better than low frequency a.c. or d.c. means it can be used as a **high-pass filter** for electrical signals. That is to say, it allows a high frequency electrical signal to pass, but blocks any low frequency signals. This is particularly useful if a small a.c. voltage is superimposed on a large d.c. voltage, and we are trying to measure the a.c. part. Figure 5.8 shows how a high pass filter is used to measure such a signal (a). The d.c. component can be filtered out using a capacitor, leaving the signal shown in Figure 5.8 (b). The sensitivity of the voltmeter can then be turned up to allow the a.c. signal to be measured accurately (c).

Figure 5.8: Filtered and amplified a.c. signal

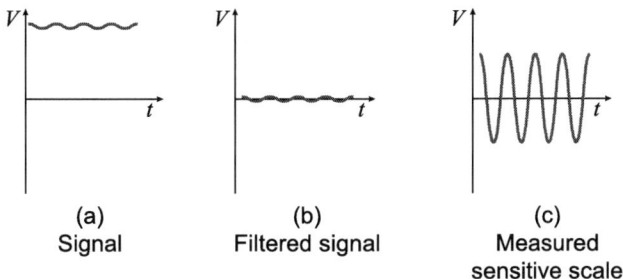

(a)	(b)	(c)
Signal	Filtered signal	Measured sensitive scale

5.5 Capacitive reactance

The exact relationship between capacitive reactance and a.c. frequency is given by the expression $X_C = \frac{1}{2\pi f C}$

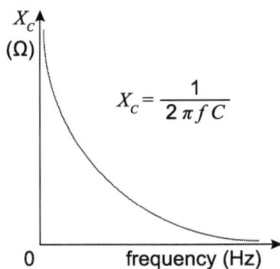

$$X_c = \frac{1}{2\pi f C}$$

A graph of X_C against $\frac{1}{f}$ is a straight line through the origin, with gradient equal to $\frac{1}{2\pi C}$.

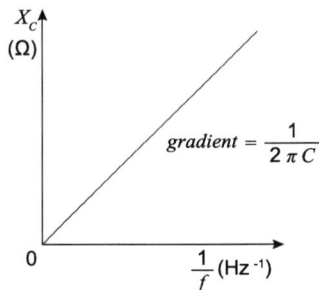

$$gradient = \frac{1}{2\pi C}$$

Example

Problem:

A 4700 μF capacitor is connected to an a.c. supply of frequency 12 Hz. The r.m.s voltage is 6.0 V. Calculate the r.m.s. current.

Solution:

$$X_C = \frac{1}{2\pi f C}$$

$$X_C = \frac{1}{2\pi \times 12 \times 4700 \times 10^{-6}}$$

$$X_C = 2.82 \ \Omega$$

$$X_C = \frac{V}{I}$$

$$2.82 = \frac{6.0}{I}$$

$$I = 2.1 \text{ A}$$

....

5.6 Summary

Summary

You should now be able to:

- sketch graphs of voltage and current against time for charging and discharging capacitors in series CR circuits;

- define the time constant of a circuit;

- carry out calculations relating the time constant of a circuit to the resistance and capacitance;

- use graphical data to determine the time constant of a circuit;

- define capacitive reactance;

- describe the response of a capacitive circuit to an a.c. signal;

- use the appropriate relationship to solve problems involving capacitive reactance, voltage and current;

- use the appropriate relationship to solve problems involving a.c. frequency, capacitance and capacitive reactance.

5.7   Extended information

Web links

There are web links available online exploring the subject further.

. . . .

5.8   Assessment

End of topic 5 test

The following test contains questions covering the work from this topic.

Go online

Q1:   The following graph shows how the charge stored by a capacitor varies with time as it charges in a CR circuit.

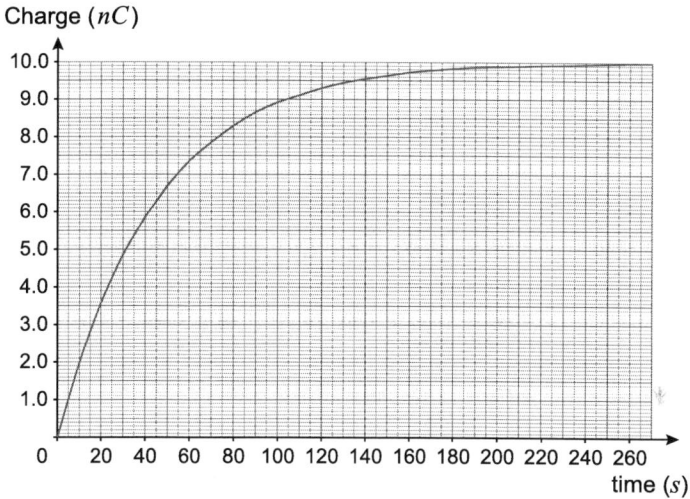

Charge (nC)

Use the graph to determine the time constant of the discharge circuit.

_____ s

. . . .

Q2: The following graph shows how the potential difference across a capacitor varies with time as it discharges in a CR circuit.

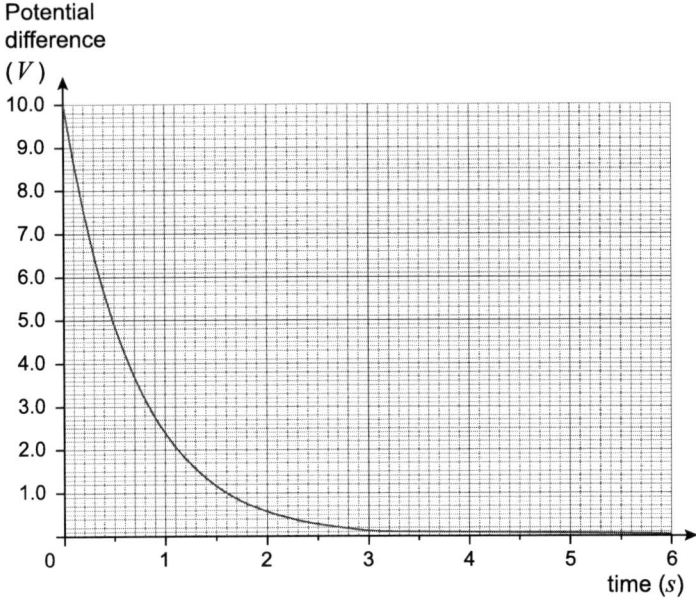

The capacitor has a capacitance of 4700 μF. Determine the resistance of the circuit.

_____ Ω

....

Q3: In a series CR circuit, a 580 nF capacitor and a 400 Ω resistor are connected to an a.c. power supply. When the frequency of the supply is 50 Hz, the r.m.s current in the circuit is 17 mA.

Calculate the r.m.s. current when the frequency of the supply is increased to 150 Hz, if the r.m.s. voltage of the power supply is kept constant.

_____ mA

....

Q4: The frequency of the output from an a.c. supply is increased.
Which graph shows how the reactance of a capacitor varies with the frequency of the supply?

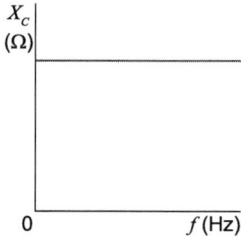

(a)

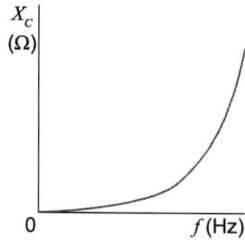

(b)

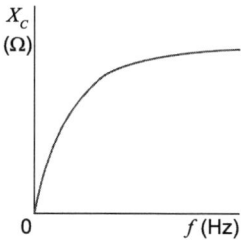

(c)

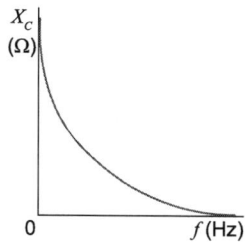

(d)

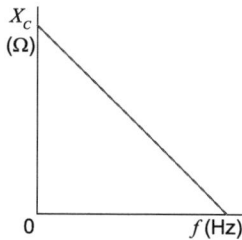

(e)

. . . .

Q5: A 180 nF capacitor is connected to 14.0 V a.c. power supply. The frequency of the a.c. supply is 5800 Hz.

1. Calculate the capacitive reactance of the capacitor.
   _ _ _ _ _ _ _ _ _ _ Ω
2. Calculate the current in the circuit.
   _ _ _ _ _ _ _ _ _ _ A

. . . .

Topic 6

Inductors (Unit 3)

Contents

Prerequisite knowledge

- *Magnetic fields (Unit 3 - Topic 4).*

- *Energy and power in an electric circuits (CfE Higher).*

- *Current and voltage in series and parallel circuits (CfE Higher).*

- *Capacitors in a.c. circuits (Unit 3 - Topic 5).*

Learning objectives

By the end of this topic you should be able to:

- *sketch graphs showing the growth and decay of current in a simple d.c. circuit containing an inductor;*

- *describe the principles of a method to illustrate the growth of current in a d.c. circuit;*

- *state that an e.m.f. is induced across a coil when the current through the coil is varying;*

- *explain the production of the induced e.m.f across a coil;*

- *explain the direction of the induced e.m.f in terms of energy;*

- *state that the inductance of an inductor is one henry if an e.m.f. of one volt is induced when the current is changing at a rate of one ampere per second;*

- *use the equation $\varepsilon = -L\frac{dI}{dt}$ and explain why a minus sign appears in this equation;*

- *state that the work done in building up the current in an inductor is stored in the magnetic field of the inductor, and that this energy is given by the equation $E = \frac{1}{2}LI^2$;*

- *calculate the maximum values of current and induced e.m.f. in a d.c. LR circuit;*

- *use the equations for inductive reactance $X_L = \frac{V}{I}$ and $X_L = 2\pi fL$;*

- *describe the response of an a.c. inductive circuit to low and high frequency signals.*

6.1 Introduction

Electromagnetic induction is the production of an induced e.m.f. in a conductor when it is present in a changing magnetic field. An airport metal detector is just one example of a modern appliance that relies on electromagnetic induction for its operation. In this topic we will investigate different ways of producing an induced current and we will look at various other applications of this effect.

We will then focus on the behaviour of inductors, which are basically coils of wire designed for use in electronic circuits. We will pay particular attention to their opposition to current flow, allowing us to contrast their behaviour to that of the capacitors we studied in the last topic.

6.2 Magnetic flux and induced e.m.f.

6.2.1 Magnetic flux and solenoids

Before we look at induction, we will first review some essential points concerning magnets and magnetic fields. An important concept is **magnetic flux**. We can visualise the magnetic flux lines to indicate the strength and direction of a magnetic field, just as we have used field lines to represent electrical or gravitational fields. The magnetic flux ϕ passing through an area A perpendicular to a uniform magnetic field of strength B is given by the equation

$$\phi = BA$$

(6.1)

. . . .

where ϕ is the flux density in T m^2 (or weber, Wb). While you do not need to remember this equation, the idea of magnetic flux is a useful one in understanding inductance.

Another idea that you should have met before is the magnetic field associated with a current-carrying coil, otherwise known as a solenoid. The magnetic field strength inside an air-filled cylindrical solenoid depends on the radius and length of the coil, and the number of turns of the coil. The direction of the magnetic field depends on the direction of the current, as shown in Figure 6.1.

Figure 6.1: Solenoids - the direction of the current (electron flow) tells us the direction of the magnetic field

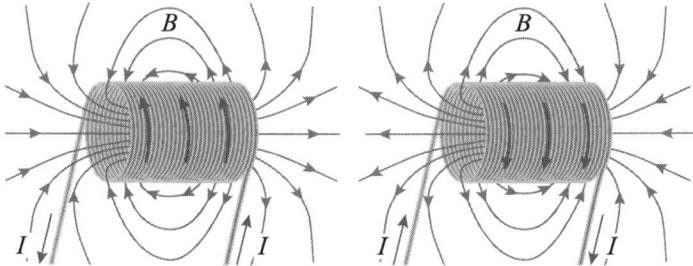

. . . .

6.2.2   Induced e.m.f. in a moving conductor

In previous topics we have studied the force exerted on a charge moving in a magnetic field, such as the charged particles making up the solar wind. We begin this topic by looking at how this force can induce an e.m.f. in a conductor.

Any metallic conductor contains 'free' electrons that are not strongly bonded to any particular atom. When an e.m.f. is applied, these electrons drift along in the conductor, this movement of charges being what we call an electric current. We have used the equation $F = IlB \sin \theta$ to calculate the force on a conductor placed in a magnetic field when a current is present.

Consider now what happens when a rod of metal is made to move in a magnetic field.

Figure 6.2: Metal rod moving at right angles to a magnetic field

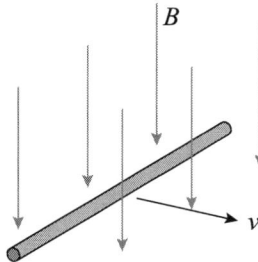

. . . .

In Figure 6.2, a magnetic field acts vertically downwards in the diagram. As the conductor is moved from left to right, each free electron is a charged particle moving at right angles to a magnetic field. The force on each electron acts out of the page in the diagram, so electrons will drift that way, leaving a net positive charge behind. Thus there is a net positive charge at the far away end of the end of the rod and a net negative charge at the other end. This means there will be a potential difference between the ends of the rod.

A force will act on the charges in a conductor whenever a conductor moves across a magnetic field. We usually state that this occurs whenever a conductor crosses magnetic flux lines, as there is no induced voltage when the conductor moves parallel to the magnetic field. If the conductor is connected to a stationary circuit, as shown in Figure 6.3, then a current I is induced in the circuit.

Figure 6.3: Current due to the induced e.m.f

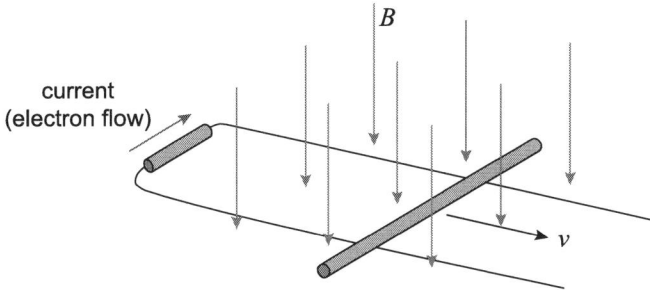

You should note that the **induced e.m.f.** occurs when there is relative motion between the magnetic field and a conductor, so we can have a stationary conductor and a moving magnetic field. An example of this is the e.m.f. induced when a magnet is moved in and out of a coil, as shown in Figure 6.4.

Figure 6.4: Induced e.m.f. causing a current to appear in a coil when the magnet is moved up and down

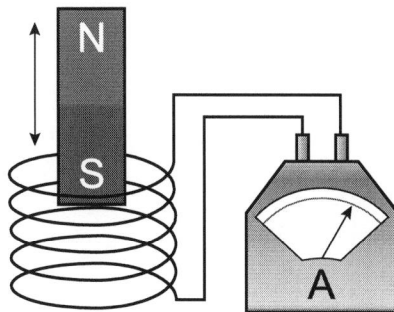

Induced e.m.f.

Go online

There is an activity available online, which allows you to investigate induced current caused by the change in magnetic field.

. . . .

As the magnet is moved in and out of the coil, the induced e.m.f. causes a current. The direction of the current changes as the direction of the magnet's movement changes. If the magnet is stationary, whether inside or outside the coil, no current is detected. The induced e.m.f. only appears when there is relative movement between the coil and the magnet.

In fact, it is the change of magnetic flux that causes the induced e.m.f., and the magnitude of the induced e.m.f. is proportional to the rate of change of magnetic flux. This means that if the magnetic field strength is changing, an e.m.f. is induced in a conductor placed in the field. So an e.m.f. can be induced by changing the strength of a magnetic field without needing to physically move a magnet or a conductor. This is the effect we will be studying in the remainder of this topic.

6.2.3 Cassette players

We have just described how a moving magnet can induce a current in a coil. Exactly the same principle is used in the playback head of a cassette player, the device people used to listen to music before mp3 players were invented. The tape in a pre-recorded cassette is magnetised, and is effectively a collection of very short bar magnets spaced along the tape. The head consists of an iron ring with a small gap, under which the tape passes. As the tape passes under the ring, the ring becomes magnetised, the direction and strength of the field in the ring constantly changes as the tape passes under it.

A coil of wire wound around the top of the ring is connected to an amplifier circuit. As the magnetic field in the ring changes, a current is induced in the coil. It is this electrical signal that is amplified and played through the speakers.

6.2.4 Faraday's law and Lenz's law

Electromagnetic induction was investigated independently by the English physicist Michael Faraday and the German physicist Heinrich Lenz in the mid-19th century. The laws which bear their names tell us the magnitude and direction of the induced e.m.f. produced by electromagnetic induction.

Faraday's law of electromagnetic induction states that the magnitude of the induced e.m.f. is proportional to the rate of change of magnetic flux through the coil or circuit.

Lenz's law states that the induced current is always in such a direction as to oppose the change that is causing it.

These two laws are summed up in the relationship

$$\varepsilon \propto -\frac{d\phi}{dt}$$

(6.2)

....

where ε is the induced e.m.f. Lenz's law is essentially a statement of conservation of energy: to induce a current, we have to put work into a system.

Looking back at Figure 6.4, Faraday's law tells us that the faster we move the magnet up and down, the larger the induced e.m.f. will be. A current around a coil produces its own magnetic field (see Figure 6.1), and Lenz's law tells us that this field will cause a force that will oppose the motion of the bar magnet towards the coil. Similarly, when the bar magnet is being removed from the coil, the induced current causes an attractive force on the bar magnet, again opposing its motion.

6.3   Eddy currents

Consider a metal disc rotating about its centre, as shown in Figure 6.5.

Figure 6.5: Rotating metal disc with a magnetic field acting on a small part

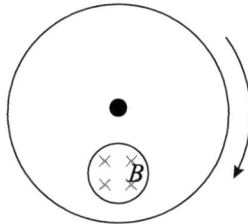

....

We will consider a magnetic field acting at right angles to the disc, but only acting over a small area. If the direction of the flux lines is into the disc, and the disc is rotating clockwise, then there will be an induced current in the region of the magnetic field. This induced current is shown in Figure 6.6.

Figure 6.6: Eddy currents inside (solid line) and outside (dashed line) the magnetic field.

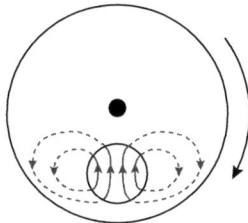

. . . .

Because the field is only acting on part of the disc, charge will be able to flow back in the regions of the disc that are outside the field (shown as dashed lines in Figure 6.6). Thus **eddy currents** are induced in the disc. Note that in the region of the field, the charge is all flowing in one direction (solid lines), and the force that acts because of this is in the opposite direction to the rotation of the disc.

6.3.1   Electromagnetic braking

Lenz's law tells us that an induced current always opposes the change that is causing it. This means that eddy currents can be used to supply **electromagnetic braking**. Consider the induced current in a localised field acting on part of a freely spinning disc, as shown in the solid lines in Figure 6.6.

The eddy currents in the part of the disc within the magnetic field cause a force to act on the disc in the opposite direction to the rotation of the disc. The currents in the opposite direction (dashed lines) are outside the field, so do not contribute a force. Thus a net force opposing the motion acts on the disc, slowing it down. This effect is used in circular saws, to bring the saw blade to rest quickly after the power is turned off. The same effect is used as the braking system in electric rapid-transit trains.

6.3.2   Induction heating

Eddy currents can lead to a large amount of energy being lost in electric motors through **induction heating**. The power dissipated when there is a current I through a resistor R is equal to $I^2 R$, so large currents can lead to a lot of energy being transferred as heat energy. In large dynamos in power stations, for example, this can make the generation of useful energy very inefficient. The laminated dynamos described earlier reduce eddy currents and hence reduce induction heating.

Induction heating is not always undesirable. In fact, it is used in circumstances where other forms of heating are impractical, such as the heat treatment of metals - welding and soldering. A piece of metal held in an electrical insulator can be heated to a very high temperature by the eddy currents. No eddy current is induced in the insulator, so it will remain cool.

Modern cookers called induction hobs work by electrical induction rather than by thermal conduction from a flame or a heating element. Therefore, such cookers require the use of a pot made of a ferromagnetic metal such as iron or stainless steel. They don't work with copper and aluminium pots.

The cooking pot is placed above a coil of copper wire that has an alternating current passing through it. This results in a changing magnetic field, which induces eddy currents in the pot, causing it to heat up. Since nothing outside the vessel is affected by the field, it is a very efficient process, only heating the pot itself. Furthermore, it is much safer since the induction cooking surface is only heated by the pot rather than by a heating element, making people less likely to receive a burn.

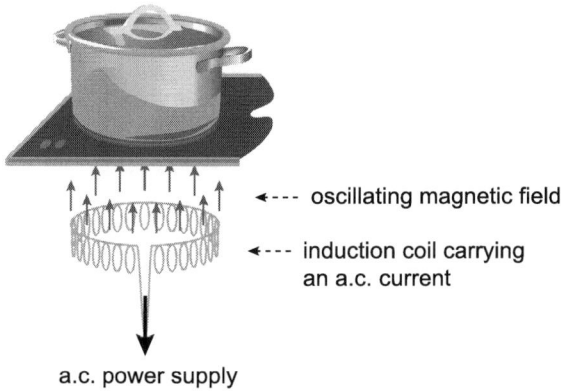

← - - - oscillating magnetic field

← - - - induction coil carrying
an a.c. current

a.c. power supply

6.3.3   Metal detectors

An airport metal detector uses eddy currents to generate a magnetic field, and it is this field that is actually detected.

Figure 6.7: Schematic of an airport metal detector

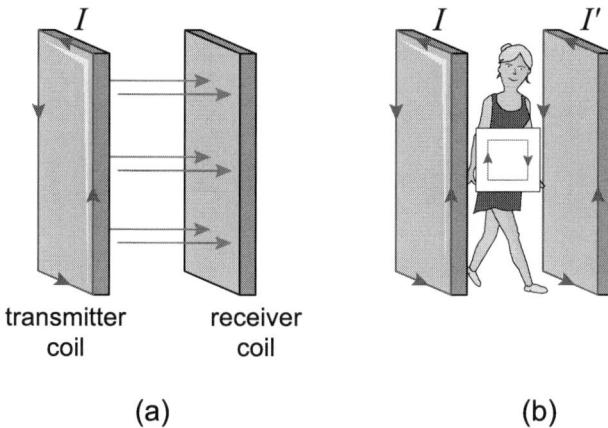

transmitter   receiver
coil   coil

(a)   (b)

Each passenger passes between two coils . The steady current I in the transmitter coil creates a magnetic field B (Figure 6.7 (a)). If a passenger walking between the coils is carrying a metal object, then eddy currents are induced in the object, and these currents in turn produce their own (moving) magnetic field. This new magnetic field induces a current I' in the receiver coil (Figure 6.7 (b)), triggering the alarm.

6.4   Inductors and self-inductance

We are now going to explore the function of inductors, which are coils of wire designed for use in electronic circuits.

6.4.1   Self-inductance

A coil (or inductor, as we shall see) in an electrical circuit can be represented by either of the symbols shown in Figure 6.8.

Figure 6.8: Circuit symbols for (a) an air-cored inductor; (b) an iron-cored inductor

(a)   (b)

Let us consider a simple circuit in which a coil of negligible resistance is connected in series to a d.c. power supply and a resistor (Figure 6.9).

Figure 6.9: Coil connected to a d.c. power supply

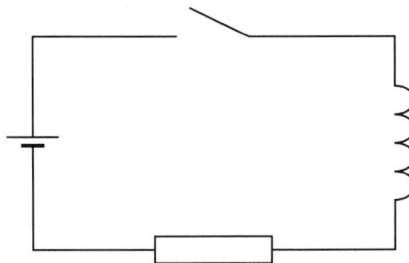

When a steady current is present, the magnetic field in and around the coil is stable. When the current changes (when the switch is opened or closed), the magnetic field changes and an e.m.f. is induced in the coil. This e.m.f. is called a self-induced e.m.f., since it is an e.m.f. induced in the coil that is caused by a change in its own magnetic field. The effect is known as **self-inductance**.

We know from Equation 6.2 that the induced e.m.f. ε is proportional to the rate of change of magnetic flux. Since the rate of change of the magnetic flux in a coil is proportional to the rate of change of current, we can state that

$$\varepsilon \propto -\frac{d\phi}{dt}$$
$$\therefore \varepsilon \propto -\frac{dI}{dt}$$

(6.3)

....

The constant of proportionality in Equation 6.3 is the inductance L of the coil. The inductance depends on the coil's size and shape, the number of turns of the coil, and whether there is any material in the centre of the coil. A coil in a circuit is called a self-inductor (or more usually just an **inductor**). The self-induced e.m.f. ε in an inductor of inductance L is given by the equation

$$\varepsilon = -L\frac{dI}{dt}$$

(6.4)

....

In Equation 6.4, $\frac{dI}{dt}$ is the rate of change of current in the inductor. The SI unit of inductance is the henry (H). From Equation 6.4 we can see that an inductor has an inductance L of 1 H if an e.m.f. of 1 V is induced in it when the current is changing at a rate of 1 A s⁻¹. Note that there is a minus sign in Equation 6.4, consistent with Lenz's law. The self-induced e.m.f. always opposes the change in current in the inductor, and for this reason is also known as the **back e.m.f.**.

6.4.2 Energy stored in an inductor

Let us return to Figure 6.9 and consider an ideal inductor - one with negligible resistance. When the switch is closed, the current in the inductor increases from zero to some final value I. Work is done by the power supply against the back e.m.f., and this work is stored in the magnetic field of the inductor. We will now find an expression to enable us to calculate how much energy E is stored in the magnetic field.

You should already be aware that if a potential difference V exists across a component in a circuit when a current I is present, then the rate P at which energy is being supplied to that component is given by the equation $P = IV$ (where P is measured in W, equivalent to J s⁻¹). If the current is varying across an inductor, then we can use Equation 6.4 and substitute for the potential difference

$$P = IV$$
$$\therefore P = I \times L\frac{dI}{dt}$$

(Since we are only concerned with the magnitude of the potential difference, we have ignored the minus sign when making this substitution.) P is the rate at which energy is being supplied, so we can substitute for $P = dE/dt$

$$\frac{dE}{dt} = LI\frac{dI}{dt}$$
$$\therefore dE = LI\,dI$$

Integrating over the limits from zero to the final current I

$$\int_0^E dE = \int_0^I LI\,dI$$
$$\therefore \int_0^E dE = L\int_0^I I\,dI$$
$$\therefore E = \frac{1}{2}LI^2$$

(6.5)

. . . .

Example

Problem:

A 2.0 H inductor is connected into a simple circuit. If a steady current of 0.80 A is present in the circuit, how much energy is stored in the magnetic field of the inductor?

Solution:

Using Equation 6.5

$$E = \frac{1}{2}LI^2$$
$$\therefore E = \frac{1}{2} \times 2 \times 0.80^2$$
$$\therefore E = 0.64\,J$$

. . . .

The energy stored in the magnetic field of an inductor can itself be a source of e.m.f. When the current is switched off, there is a change in current so a self-induced e.m.f. will appear across the inductor opposing the change in current. The energy used to create this e.m.f. comes from the energy that has been stored in the magnetic field.

Quiz: Self-inductance

Q1: A potential difference can be induced between the ends of a metal wire when it is

a) moved parallel to a magnetic field.
b) moved across a magnetic field.
c) stationary in a magnetic field.
d) stationary outside a solenoid.
e) stationary inside a solenoid.

Go online

....

Q2: Lenz's law states that

a) the induced e.m.f. in a circuit is proportional to the rate of change of magnetic flux through the circuit.
b) the magnetic field in a solenoid is proportional to the current through it.
c) magnetic flux is equal to the field strength times the area through which the flux lines are passing.
d) the induced current is always in such a direction as to oppose the change that is causing it.
e) the induced current is proportional to the magnetic field strength.

....

Q3: The current in an inductor is changing at a rate of 0.072 A s^{-1}, producing a back e.m.f. of 0.021 V.

What is the inductance of the inductor?

a) 0.0015 H
b) 0.29 H
c) 3.4 H
d) 4.1 H
e) 670 H

....

Q4: The steady current through a 0.050 H inductor is 200 mA.

What is the self-induced e.m.f. in the inductor?

a) 0 V
b) -0.010 V
c) -0.25 V
d) -4.0 V
e) -100 V

....

Q5: Which **one** of the following statements is true?

a) When the current through an inductor is constant, there is no energy stored in the inductor.
b) Faraday's law does not apply to self-inductance.
c) A back e.m.f. is produced whenever there is a current through an inductor.
d) The self-induced e.m.f. in an inductor always opposes the change in current that is causing it.
e) The principle of conservation of energy does not apply to inductors.

. . . .

Q6: How much energy is stored in the magnetic field of a 4.0 H inductor when the current through the inductor is 300 mA?

a) 0.18 J
b) 0.36 J
c) 0.60 J
d) 0.72 J
e) 2.4 J

. . . .

Q7: An inductor stores 0.24 J of energy in its magnetic field when a steady current of 0.75 A is present. If the resistance of the inductor can be ignored, calculate the inductance of the inductor.

a) 0.10 H
b) 0.41 H
c) 0.43 H
d) 0.85 H
e) 1.2 H

. . . .

6.5   Inductors in d.c. circuits

We can connect an inductor into a circuit in the same way as we would connect a resistor or a capacitor. We will now investigate how an inductor behaves when it is used as a component in a circuit.

We will begin by looking at a d.c. circuit, where we have an inductor connected in series to a resistor and a power supply such as a battery. After that we will replace the battery by an a.c. supply to investigate the response of an inductive circuit to an alternating current. We will compare the responses of inductive and capacitative circuits to an a.c. signal.

6.5.1   Growth and decay of current

Consider a simple circuit with an inductor of inductance L and negligible resistance connected in series to a resistor of resistance R, an ammeter of negligible resistance, and a d.c. power supply of e.m.f. E with negligible internal resistance.

The circuit (often called simply an LR circuit) is shown in Figure 6.10.

Figure 6.10: d.c. circuit with resistor and inductor in series

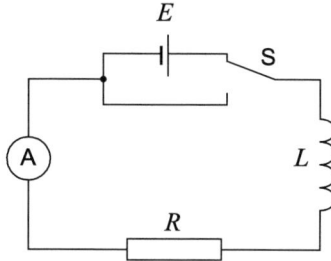

When the switch S is connected to the power supply, charge flows through the resistor and inductor, with the ammeter measuring the current. In the time taken for the current to rise from zero to its final value, the current through the inductor is changing, so a back e.m.f. is induced, which (by Lenz's law) opposes the increase in current. The rise time for the current to reach its final value in an inductive circuit will therefore be longer than it is in a non-inductive circuit.

A student could use a stopwatch to measure the time and the current could be noted from the ammeter at regular time intervals. A graph of current against time would be obtained as shown in Figure 6.11

The final, steady value of the current is given by Ohm's law, $I = {}^{E}\!/_{R}$, and so does not depend on the value of L. This should not be surprising, since when the current is at a steady value, there will be no induced back e.m.f.

Figure 6.11: Growth of current in a simple inductive circuit

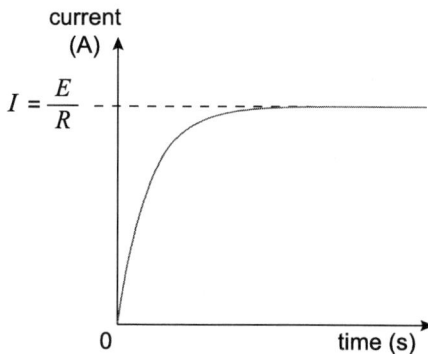

When the switch S in Figure 6.10 is switched to the down position, the power supply is no longer connected in the circuit, and the current drops from a value I to zero. Once again, the change in current produces a back e.m.f. that opposes the change. The upshot of this is that the current takes longer to decay than it would in a non-inductive circuit. This is shown in Figure 6.12.

Figure 6.12: Decay of current in a simple inductive circuit

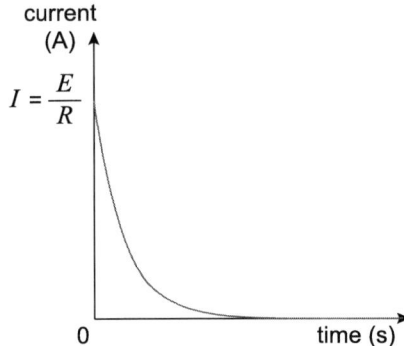

An example of the way current varies in an inductive circuit is the fact that a neon bulb connected to a battery can be lit, even although such a bulb requires a large p.d. across it. Consider the circuit shown in Figure 6.13.

Figure 6.13: Neon bulb connected to an inductive circuit

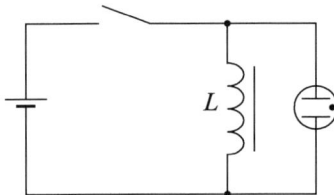

The power supply is a 1.5 V battery. We will consider an inductor that has an inductance L and a resistance R, which is connected in parallel to a neon bulb. The bulb acts like a capacitor in the circuit. Unless a sufficiently high p.d. is applied across it, the bulb acts like a break in the circuit. If the p.d. is high enough, the 'capacitor' breaks down, and charge flows between its terminals, causing the bulb to light up.

If the switch in Figure 6.13 is closed, current appears through the inductor but not through the bulb, since the p.d. across it is too low. The current rises as shown in Figure 6.11, reaching a steady value of $I = \frac{E}{R}$, where E is the e.m.f. of the battery (1.5 V) and R is the resistance of the inductor. The energy stored in the inductor is equal to $\frac{1}{2}LI^2$.

Opening the switch means there is a change in current through the inductor, and hence a back e.m.f. Charge cannot now flow around the left hand side of the circuit. It can only flow across the neon bulb, causing the bulb to flash.

6.5.2 Back e.m.f.

In the previous topic, we stated that the back e.m.f.

$$\varepsilon$$

induced in an inductor of inductance L is given by the equation

$$\varepsilon = -L\frac{dI}{dt}$$

Figure 6.11 shows that the rate of change of current in an LR circuit such as in Figure 6.10 is greatest just after the switch is moved to the battery side, so this is when the back e.m.f. will also be at its largest value. As the current increases, the rate of change decreases, and hence the back e.m.f. also decreases.

The sum of e.m.f.s around a circuit loop is equal to the sum of potential differences around the loop. So, at the instant when the switch is moved to the battery side, the current in the resistor is zero. This means the back e.m.f. must be equal in magnitude to E. As the current grows (and the *rate of change* of current *decreases*), the p.d. across the resistor increases and the back e.m.f. must decrease. Thus we have a maximum value for the back e.m.f., which is

$$\varepsilon_{\max} = -E$$

(6.6)

. . . .

The minus sign appears as the back e.m.f. is, by its very nature, in the opposite direction to E.

Example

Problem:

The circuit in Figure 6.14 contains an ideal (zero resistance) inductor of inductance 2.5 H, a 500 Ω resistor and a battery of e.m.f. 1.5 V and negligible internal resistance.

Figure 6.14: Inductor, resistor and battery in series

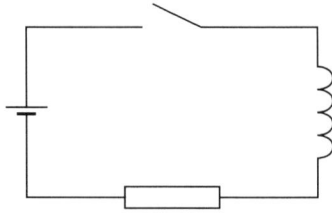

(a) What is the maximum self-induced e.m.f. in the inductor?

(b) At the instant when the back e.m.f. is 0.64 V, at what rate is the current changing in the circuit?

Solution:

Solution

(a) The maximum self-induced e.m.f. is equal and opposite to the e.m.f. of the battery, which is -1.5 V.

(b) Using the equation for back e.m.f.

$$\varepsilon = -L\frac{dI}{dt}$$
$$\therefore \frac{dI}{dt} = -\frac{\varepsilon}{L}$$
$$\therefore \frac{dI}{dt} = -\frac{-0.64}{2.5}$$
$$\therefore \frac{dI}{dt} = 0.26\,\mathrm{A\,s^{-1}}$$

Quiz: Inductors in d.c. circuits

Go online

Q8: A 120 mH inductor is connected in series to a battery of e.m.f. 1.5 V and negligible internal resistance, and a 60 Ω resistor. What is the maximum current in this circuit?

a) 12.5 mA
b) 25 mA
c) 180 mA
d) 3 A
e) 12.5 A

Q9: In the circuit in the previous question, what is the maximum potential difference across the inductor?

a) 0 V
b) 25 mV
c) 120 mV
d) 1.5 V
e) 12.5 V

. . . .

Q10: A series circuit consists of a 9.0 V d.c. power supply, a 2.5 H inductor and a 1.0 kΩ resistor. What is the magnitude of the back e.m.f. when a steady current of 9.0 mA is present?

a) 0 V
b) 22.5 mV
c) 2.5 V
d) 3.6 V
e) 9.0 V

. . . .

Q11: The circuit shown is used to measure the growth of current in an inductor.

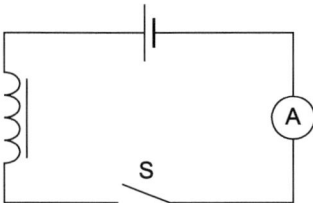

What other piece of apparatus is needed as well as the circuit above?

a) data capture device
b) digital ammeter
c) low value inductor
d) stopwatch
e) analogue voltmeter

. . . .

6.6   Inductors in a.c. circuits

You will recall from the last topic that when an a.c. current is present a resistor behaves in exactly the same way as it does for a d.c. current. Meanwhile, a capacitor was found to oppose high frequency a.c. less than low frequency. That is, capacitive reactance was found to be inversely proportional to the frequency of an alternating current.

Inductors also oppose a.c. current, and we can define an inductor's reactance as

$$X_L = \frac{V}{I}$$

Inductive reactance, like capacitive reactance, is measure in ohms (Ω).

We have mentioned Lenz's law several times in this topic. The induced e.m.f. always opposes the change that is causing it. So an ideal inductor does not oppose d.c. current. So long as the current does not vary with time, an ideal inductor offers no opposition to current. However, in a.c. circuits the current and the associated magnetic field are continually changing. As the a.c. supply's frequency increases, the rate of change of current increases. So the self-induced back e.m.f. increases and therefore the inductive reactance increases. This makes the current decrease.

Let us now explore the exact relationship between inductive reactance and an a.c. supply's frequency. Assume the inductor has negligible resistance.

Go online

The inductor and a.c.

There is an online activity where you can find out how the frequency of the a.c. supply affects the current.

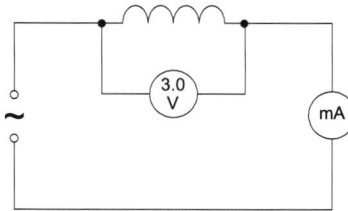

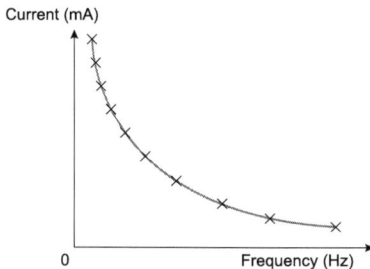

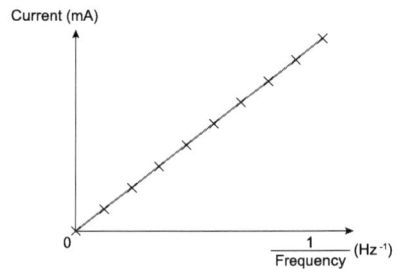

. . . .

Figure 6.15: Simple a.c. inductive circuit

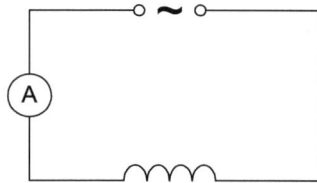

. . . .

We have found that for an inductor the current is inversely proportional to the frequency. Since $X_L = \frac{V}{I}$, this means the inductive reactance must be directly proportional to the frequency of the supply.

The relationship is given by the following expression:

$$X_L = 2\pi f L$$

Since inductive reactance increases with frequency, an inductor is a good 'low-pass' filter. An inductor allows low-frequency and d.c. signals to pass but offers high 'resistance' to high-frequency signals. An inductor can be used to smooth a signal by removing high-frequency noise and spikes in the signal.

The frequency response of a resistor, capacitor and inductor can be summarised by the following graphs.

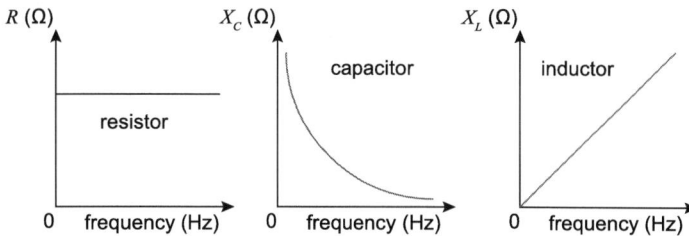

Go online

Quiz: a.c. circuits

Q12: Which of the following describes the relationship between reactance X and frequency f in an a.c. inductive circuit?

a) $X \propto {}^{1}/_{f}$

b) $X \propto {}^{1}/_{f^2}$

c) $X \propto f$

d) $X \propto f^2$

e) $X \propto \sqrt{f}$

....

Q13: Which *one* of the following statements is true?

a) An inductor can be used to filter out the d.c. component of a signal.

b) The inductance of an inductor is inversely proportional to the frequency of the supply.

c) An inductor is often used to filter out low frequency signals and allow only high frequency signals to pass through.

d) The reactance of a capacitor is proportional to the frequency of the a.c. current.

e) An inductor can smooth a signal by filtering out high frequency noise and spikes.

....

Q14: An experiment is carried out to investigate how the current varies with frequency in an inductive circuit. The results of such an experiment show that

a) $I \propto {}^{1}/_{f}$

b) $I \propto {}^{1}/_{f^2}$

c) $I \propto f$

d) $I \propto f^2$

e) $I \propto \sqrt{f}$

....

6.7   Summary

In this topic we have seen that an e.m.f. can be induced in a conductor in a magnetic field when the magnetic flux changes. Thus an e.m.f. can be induced when a conductor moves across a magnetic field; when a magnet moves near to a stationary conductor; or when the strength of a magnetic field changes.

We have found out that a coil in a circuit is called a self-inductor or just an inductor. Since work is done against the back e.m.f. in establishing a current in an inductor, there is energy stored in its magnetic field whilst a current is present in an inductor.

We then studied the behaviour of inductors in simple d.c. and a.c. circuits. We have seen that an ideal (non-resistive) inductor does not have any effect on a steady d.c. current. When the current through an inductor is changing, the induced e.m.f. acts to oppose the change. Inductive reactance was shown to be proportional to the frequency of an a.c. supply, meaning that inductors are good at filtering out high frequency signals and allowing only low frequency signals to pass through.

Summary

You should now be able to:

- sketch graphs showing the growth and decay of current in a simple d.c. circuit containing an inductor;

- describe the principles of a method to illustrate the growth of current in a d.c. circuit;

- state that an e.m.f. is induced across a coil when the current through the coil is varying;

- explain the production of the induced e.m.f across a coil;

- explain the direction of the induced e.m.f in terms of energy;

- state that the inductance of an inductor is one henry if an e.m.f. of one volt is induced when the current is changing at a rate of one ampere per second;

- use the equation $\varepsilon = -L\frac{dI}{dt}$ and explain why a minus sign appears in this equation;

- state that the work done in building up the current in an inductor is stored in the magnetic field of the inductor, and that this energy is given by the equation $E = \frac{1}{2}LI^2$;

- calculate the maximum values of current and induced e.m.f. in a d.c. LR circuit;

- use the equations for inductive reactance $X_L = \frac{V}{I}$ and $X_L = 2\pi f L$;

- describe the response of an a.c. inductive circuit to low and high frequency signals.

6.8 Extended information

6.8.1 Levitation of superconductors

Another effect of eddy currents is the levitation of a superconductor in a magnetic field.

To explain this effect, we first need to understand a little about superconductivity. This effect, first observed by the Dutch physicist H. K. Onnes in 1911, is one that occurs at very low temperatures, at which certain metals and compounds have effectively zero electrical resistance. For many years, superconductivity could only be observed in materials cooled below the boiling point of helium, which is 4 K. In recent years, huge worldwide research activity has resulted in the development of compounds that can remain superconducting up to the boiling point of nitrogen (77 K).

As well as exhibiting zero electrical resistance, a superconducting material also has interesting magnetic properties. A piece of superconductor placed in a magnetic field will distort the field lines, so that the magnetic field inside the superconductor is zero.

Figure 6.16 shows a metal sphere and a superconducting sphere in a uniform magnetic field.

Figure 6.16: (a) Uniform magnetic field; (b) iron sphere placed in the field; (c) superconducting sphere placed in the field

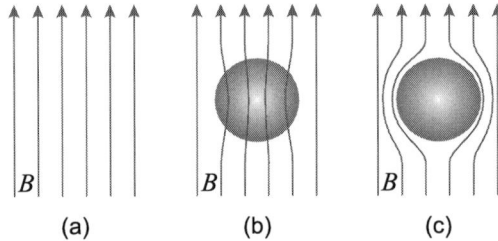

We are considering a uniform magnetic field B, shown in Figure 6.16 (a). The magnetic field strength inside an iron sphere, for example, (Figure 6.16 (b)) is enhanced. On the other hand, a superconductor placed in the field (Figure 6.16 (c)) distorts the field so that no field lines can enter it. The magnetic field inside the superconductor is therefore zero.

We can explain this phenomenon in terms of eddy currents. When the superconductor is moved into the magnetic field, eddy currents are induced on its surface. Lenz's law states that these currents will create a magnetic field opposing the external field. Since there is no electrical resistance in a superconductor, the eddy currents will continue even when the superconductor is stationary in the field. The magnetic field due to the eddy currents is in the opposite direction to the external magnetic field, with the result that the external magnetic field cannot penetrate into the superconductor. Since magnetic field lines cannot be broken, the lines must continue outside the superconductor.

We can see this effect by placing a superconductor in the field of a permanent magnet. Let us first think about what happens if we place a piece of iron near a permanent magnet. The magnetic domains within the piece of iron arrange themselves in the direction of the magnetic field lines, and the resulting attractive force draws the piece of iron towards the magnet.

The opposite happens to a piece of superconductor. Lenz's law means that the magnetic field due to the eddy currents in the superconductor opposes the field due to the permanent magnet, and a repulsive force exists between the two. We can observe magnetic levitation if we position the superconductor above the magnet, as the force of gravity acting down on the magnet can be balanced by the magnetic repulsion acting upwards.

Levitating superconductor

Go online

At this stage there is a video clip which shows a demonstration of magnetic levitation. A piece of superconductor, cooled with liquid nitrogen, is suspended above a permanent magnet.

. . . .

6.8.2 Web links

Web links

There are web links available online exploring the subject further.

. . . .

6.9 Assessment

End of topic 6 test

The following test contains questions covering the work from this topic.

Go online

Q15: The current through a 0.55 H inductor is changing at a rate of 15 A s^{-1}.

Calculate the magnitude of the e.m.f. induced in the inductor. (Do NOT include a minus sign in your answer.)

_____ V

. . . .

Q16: A 4.5 H inductor of negligible resistance is connected to a circuit in which the steady current is 460 mA.

Calculate the energy stored in the magnetic field of the inductor.

_____ J

. . . .

Q17: Which of the following is equivalent to one henry?

a) 1 A V s^{-1}
b) 1 V A^{-1} s^{-1}
c) 1 V s A^{-1}
d) 1 A V^{-1} s^{-1}

. . . .

Q18: Consider the circuit below, in which a variable resistor R and an inductor L of inductance 1.5 H are connected in series to a 3.0 V battery of zero internal resistance.

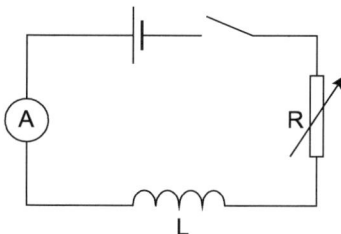

The variable resistor is changed from 76 Ω to 25 Ω over a time period of 2.5 s.

Calculate the average back e.m.f. across the inductor whilst the resistance is being changed. (Do NOT include a minus sign in your answer.)

_____ V

. . . .

Q19: A resistor R = 14 Ω and an inductor L = 580 mH are connected to a power supply as shown below.

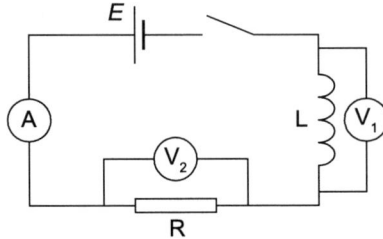

A short time after the switch is closed, the current in the circuit has reached a steady value, and the energy stored in the inductor is 0.10 J.

Calculate the e.m.f. of the power supply.

_____ V

. . . .

Q20: The circuit below shows a 12 V power supply connected to a 1.7 H inductor and a 32 Ω resistor.

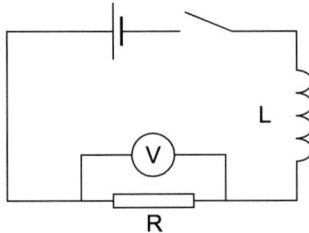

Calculate the potential difference measured by the voltmeter when the current through the resistor is changing at a rate of 2.1 A s^{-1}.

_____ V

. . . .

Q21: In the circuit shown below, the voltmeters V_1 and V_2 measure the potential difference across an inductor L and a resistor R respectively.

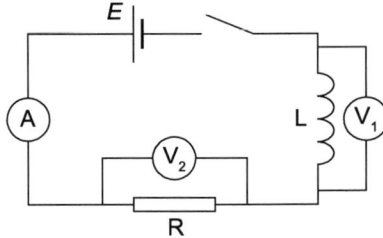

The battery has e.m.f. E.
$L = 620$ mH and $R = 14 \, \Omega$.

1. The maximum potential difference in V recorded on the voltmeter V_2 after the switch is closed is 2.6 V.
   State the e.m.f. E of the battery.
   _____ V

2. After the switch has been closed for several seconds, state the value of the potential difference measured by voltmeter V_1.
   _____ V

3. Calculate the maximum current recorded on the ammeter A after the switch is closed.
   _____ A

....

Q22: Consider the circuit shown below, in which an inductor L and resistor R are connected in series to a 1.5 V battery of negligible internal resistance.

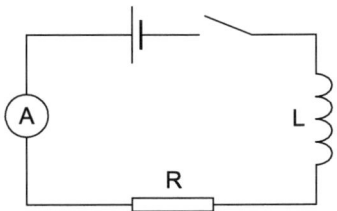

L has value 340 mH and the resistance R is 35 Ω.

1. At one instant after the switch is closed, the current in the circuit is changing at a rate of 1.9 A s^{-1}.
   Calculate the back e.m.f. at this instant. (Do NOT include a minus sign in your answer.)
   _____ V

2. Calculate the maximum current through the inductor.

    ---------- A

3. Calculate how much energy is stored in the magnetic field of the inductor when the current reaches a steady value.

    ---------- J

. . . .

Q23: Consider the circuit below, in which an inductor is connected to a 7.0 V battery of negligible internal resistance.

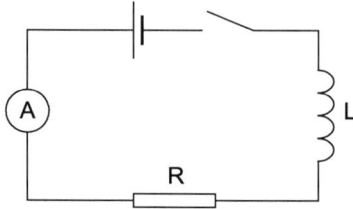

The resistance R is 40 Ω.

1. At the instant the switch is closed, the current in the circuit is changing at a rate of 60 A s^{-1}.

   Calculate the inductance L.

    ---------- H

2. Calculate the maximum current in the circuit.

    ---------- A

3. Calculate the energy store

    ---------- J

   d in the inductor when the current has reached its maximum value.

. . . .

Q24: Consider the circuit below.

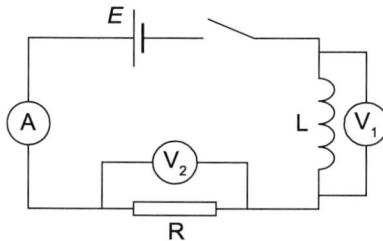

The battery has e.m.f. 2.8 V, and is connected to an inductor L = 320 mH and a resistor R = 16 Ω. The voltmeters V_1 and V_2 measure the potential difference across the inductor and the resistor respectively.

1. Calculate the maximum potential difference recorded on the voltmeter V_1 after the switch is closed.

   _____ V

2. Calculate the maximum potential difference recorded on the voltmeter V_2 after the switch is closed.

   _____ V

3. Calculate the energy stored in the magnetic field of the inductor when the current has reached a steady value.

   _____ J

. . . .

Topic 7

Electromagnetic radiation (Unit 3)

Contents

Prerequisite knowledge

- *Wave properties (Unit 2 - Topic 5).*

- *Electromagnetic waves (Unit 2 - Topic 9).*

Learning objectives

By the end of this topic you should be able to:

- *state that the similarities between electricity and magnetism led to their unification i.e. the discovery that they are really manifestations of a single electromagnetic force;*

- *state that electromagnetic radiation exhibits wave properties i.e. electromagnetic radiation reflects, refracts, diffracts and undergoes interference;*

- *describe electromagnetic radiation as a transverse wave which has both electric and magnetic field components which oscillate in phase perpendicular to each other and the direction of energy propagation;*

- *carry out calculations using $c = \frac{1}{\sqrt{\varepsilon_0 \mu_0}}$.*

7.1   Introduction

We have seen that electric currents exert forces on magnets and that time-varying magnetic fields can induce electric currents. Until the 1860s, they were thought to be unrelated. We are now going to look at the work of James Clerk Maxwell, who recognised the similarities between electricity and magnetism and developed his theory of a single electromagnetic force.

7.2   The unification of electricity and magnetism

You will be aware that the four fundamental forces of nature are gravitational, electromagnetic, and the strong and the weak nuclear forces. Theoretical physicists currently favour the idea that these four forces are actually just different manifestations of the same force. That is to say, there is only one fundamental force, and we perceive it to be acting in four different ways. One of the biggest challenges in theoretical physics is to find a Grand Unified Theory (GUT) which will unite these forces, showing that at extremely short distances, for extremely high energy particles, the four forces become one.

The Scottish physicist James Clerk Maxwell was the first to successfully unify two of these forces. His theory on electromagnetism showed that electricity and magnetism could be unified. Theoretical physicists have subsequently shown that the electromagnetic and the weak forces can be combined as an 'electroweak' force when acting over very short distances. However, this is only the case at the high energies explored in particle collisions at CERN and other laboratories. Unfortunately, it is impossible at present to study high enough energies to directly explore the unification of the other forces, but it is thought that such conditions would have existed in the early universe, almost immediately after the big bang. Instead, physicists must look for the consequences of grand unification at lower energies. Such consequences include supersymmetry, which is a theory that predicts a partner particle for each particle in the Standard Model.

7.3   The wave properties of em radiation

You will recall from Unit 2 - Topic 9 that electromagnetic waves such as light are made up of oscillating electric and magnetic fields. For simplicity, diagrams often only show the oscillating electric field, but it is important to remember that an electromagnetic wave has both electric and magnetic field components which oscillate in phase, perpendicular to each other and to the direction of energy propagation.

Figure 7.1: Electromagnetic wave

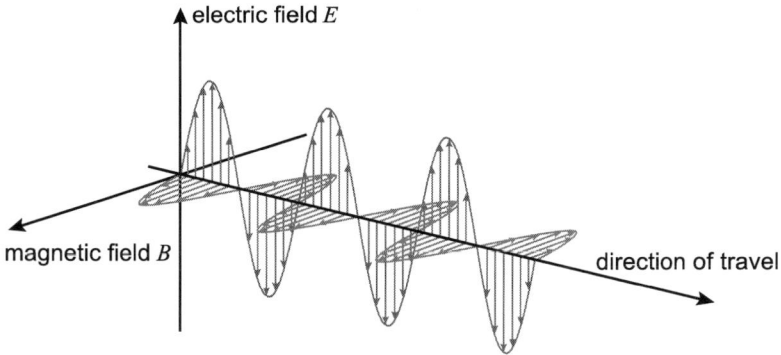

Electromagnetic wave

At this stage there is an online activity which shows a polarised electromagnetic wave that propagates in a positive x direction and explores the electric and magnetic fields.

Go online

Maxwell's theory of electromagnetism was particularly remarkable since he predicted electromagnetic waves in terms of oscillating electric and magnetic fields, long before there was any experimental evidence for them. He showed that the speed of an electromagnetic wave in a vacuum is the same as the speed of light in free space and he predicted that light is just one form of an electromagnetic wave.

In 1887, the German physicist Heinrich Hertz showed electrical oscillations give rise to transverse waves, verifying the existence of electromagnetic waves travelling at the speed of light. The waves he discovered are known now as radio waves. Bluetooth is just one example of technology we now rely upon that uses radio waves. It uses short wavelength radio waves to allow devices to communicate wirelessly. An example is a cordless telephone, which has one Bluetooth transmitter in the base and another in the handset. A computer communicating with a wireless printer, mouse or keyboard is another.

All electromagnetic radiation exhibits wave properties as it transfers energy through space. All electromagnetic radiation reflects, refracts, diffracts and undergoes interference.

7.4   Permittivity, permeability and the speed of light

Maxwell derived the equation

$$c = \frac{1}{\sqrt{\varepsilon_0 \mu_0}}$$

where
c is the speed of light in free space in ms^{-1};
ε_0 is the permittivity of free space in C^2 N^{-1} m^{-2} or F m^{-1};
μ_0 is the permeability of free space in T m A^{-1} or H m^{-1}.

Using this relationship, and the values $\varepsilon_0 = 8.85 \times 10^{-12}$ C^2 N^{-1} m^{-2} and $\mu_0 = 4\pi \times 10^{-7}$ T m A^{-1}, gives

$$\frac{1}{\sqrt{\varepsilon_0 \mu_0}} = \frac{1}{\sqrt{8.85 \times 10^{-12} \times 4\pi \times 10^{-7}}}$$
$$= 3 \times 10^8$$

Also the dimensions of $\varepsilon_0 \mu_0$ are C^2 N^{-1} m^{-2} × T m A^{-1} and since 1 C = 1 A s and 1 T = 1 N A^{-1} m^{-1}, it can be seen that the dimensions of $\frac{1}{\sqrt{\varepsilon_0 \mu_0}}$ are m s^{-1}. This shows that light is propagated as an electromagnetic wave.

In October 1983 the metre was defined as 'that distance travelled by light, in a vacuum, in a time interval of $\frac{1}{299,792,458}$ seconds'. This means that the speed of light is now a fundamental constant of physics with a value

$$c = 299,792,458 \text{ m s}^{-1}$$

7.5 Summary

Summary

You should now be able to:

- state that the similarities between electricity and magnetism led to their unification i.e. the discovery that they are really manifestations of a single electromagnetic force;

- state that electromagnetic radiation exhibits wave properties i.e. electromagnetic radiation reflects, refracts, diffracts and undergoes interference;

- describe electromagnetic radiation as a transverse wave which has both electric and magnetic field components which oscillate in phase perpendicular to each other and the direction of energy propagation;

- carry out calculations using $c = \frac{1}{\sqrt{\varepsilon_0 \mu_0}}$.

7.6 Extended information

Web links

There are web links available online exploring the subject further.

. . . .

7.7 Assessment

End of topic 7 test

The following test contains questions covering the work from this topic.

Go online

Q1: ε_0 is the symbol for the _____ of free space.

1. permeability
2. permittivity

. . . .

Q2: Electromagnetic waves are _____ .

1. longitudinal
2. transverse

. . . .

Q3: Electricity and magnetism can be _____ under one theory called electromagnetism.

....

Q4: What is the correct relationship between c, ε_0 and μ_0?

a) $c = \frac{1}{\varepsilon_0 \mu_0}$

b) $c = (\varepsilon_0 \mu_0)^2$

c) $c = \frac{1}{(\varepsilon_0 \mu_0)^2}$

d) $c = \frac{1}{\sqrt{\varepsilon_0 \mu_0}}$

e) $c = \sqrt{\varepsilon_0 \mu_0}$

....

Q5: A student carries out an experiment to determine the permittivity of free space.

It is measured to be 7.7×10^{-12} F m^{-1}.

Use this result and the speed of light in vacuum to determine the permeability of free space.

_____ H m^{-1}

....

Topic 8

End of unit 3 test

End of unit 3 test

Go online

DATA SHEET

Common Physical Quantities

The following data should be used when required:

Quantity	Symbol	Value
Charge on electron	e	-1.60×10^{-19} C
Mass of proton	m_p	1.67×10^{-27} kg
Permittivity of free space	ε_0	8.85×10^{-12} F m^{-1}
Permeability of free space	μ_0	$4\pi \times 10^{-7}$ H m^{-1}

Q1: Two point charges A (+5.95 mC) and B (+7.55 mC) are placed 1.42 m apart.

1. Calculate the magnitude of the Coulomb force that exists between A and B.
   _ _ _ _ _ _ _ _ _ _ N

2. Calculate the magnitude of the Coulomb force acting on a -1.15 mC charge placed at the midpoint of AB.
   _ _ _ _ _ _ _ _ _ _ N

· ·

Q2: A long straight wire carries a steady current I_1.

Calculate the magnetic induction at a perpendicular distance 48 mm from the wire when the current $I_1 = 1.5$ A.

_ _ _ _ _ _ _ _ _ _ T

· ·

Q3: An ion carrying charge 2e is accelerated from rest through a potential of 2.5×10^6 V, emerging with a velocity of 5.6×10^6 m s^{-1}.

Calculate the mass of the ion.

_ _ _ _ _ _ _ _ _ _ kg

· ·

Q4: Calculate the magnitude of the force on a horizontal conductor 20 cm long, carrying a current of 7.5 A, when it is placed in a magnetic field of magnitude 5.0 T acting at 33° the wire's length.

_ _ _ _ _ _ _ _ _ _ N

· ·

Q5: Consider a capacitor connected in series to an a.c. power supply.

Which one of the following graphs correctly shows how the current in the circuit varies with the frequency of the a.c. supply?

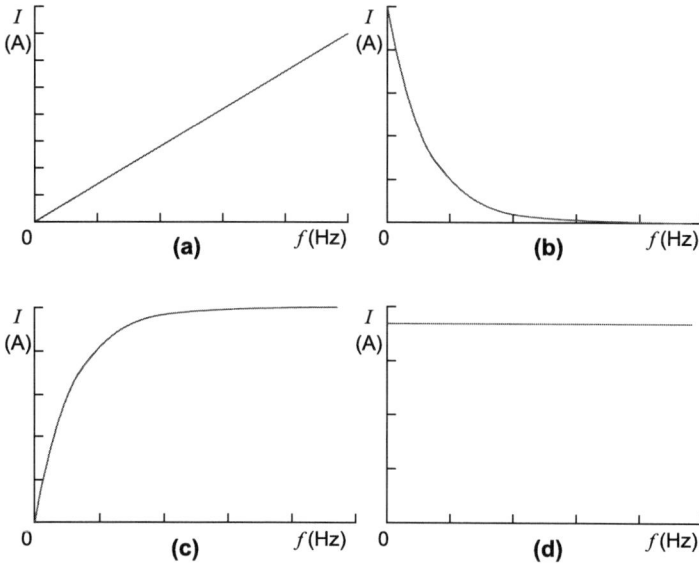

Q6: A 200 nF capacitor is connected to 1.0 V a.c. power supply. The frequency of the a.c. supply is 4600 Hz.

Calculate the capacitive reactance of the capacitor.

_____ Ω

Q7: Consider the circuit below, in which an ideal inductor L is connected in series to a resistor R and a battery of e.m.f. 6.0 V and zero internal resistance.

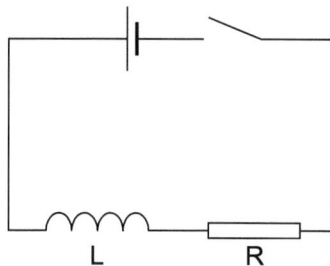

The value of L is 280 mH and the resistance R is 36 Ω.

1. Calculate the initial rate of growth of current in the circuit at the instant the switch is closed.

   _____ A s^{-1}

2. Calculate the energy stored in the magnetic field of the inductor once the current has reached a steady value.

   _____ J

. . . .

Q8: The circuit below shows a 12 V power supply connected to a 1.5 H inductor and a 36 Ω resistor.

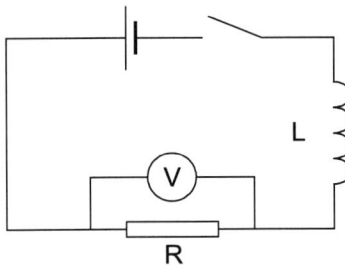

Calculate the potential difference measured by the voltmeter when the current is changing at a rate of 3.3 A s^{-1}.

_____ V

. . . .

Q9: In the circuit shown below, the voltmeters V_1 and V_2 measure the potential difference across an inductor L and a resistor R respectively.

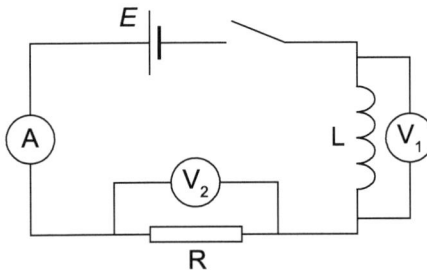

The battery has e.m.f. E. The inductor has an inductance of 660 mH and the resistance of the resistor is 14 Ω.

1. The maximum potential difference in V recorded on the voltmeter V_2 after the switch is closed is 3.2 V.
   State the e.m.f. E of the battery.
   ---------- V

2. After the switch has been closed for several seconds, state the value of the potential difference measured by voltmeter V_1.
   ---------- V

3. Calculate the maximum current recorded on the ammeter A after the switch is closed.
   ---------- A

. . . .

Q10: Consider the circuit below, in which an inductor is connected to a 8.00 V battery of negligible internal resistance.

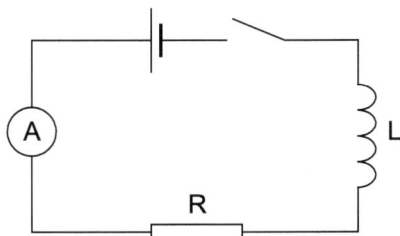

The resistance R is 40.0 Ω.

1. At the instant the switch is closed, the current in the circuit is changing at a rate of 60.0 A s^{-1}.
   Calculate the inductance L.
   ---------- H

2. Calculate the maximum current in the circuit.
   ---------- A

3. Calculate the energy stored in the inductor when the current has reached its maximum value.
   ---------- J

. . . .

Q11: Consider an inductor connected in series to an a.c. power supply.

Which one of the following graphs correctly shows how the current in the inductor varies with the frequency of the a.c. supply?

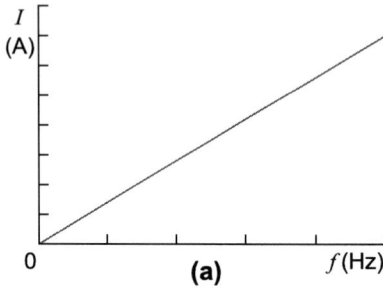

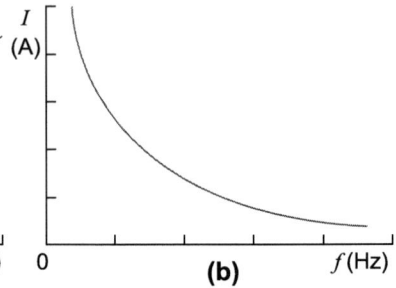

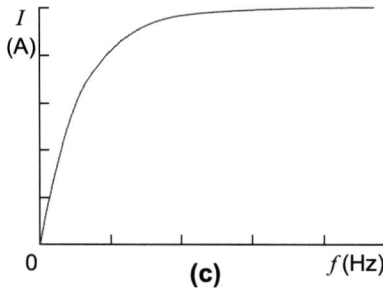

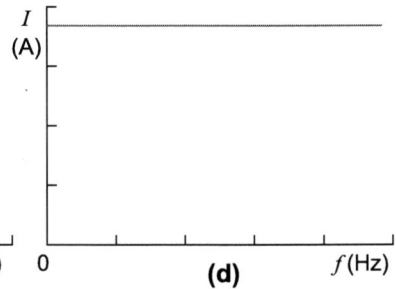

Q12: A 150 μF capacitor is connected in series with a 500Ω resistor to a 6.00 V battery. Calculate the time taken for the voltage across the capacitor to increase from 0.00 V to 3.78V.

time = _____ s

Topic 9

Initial planning, using equipment and recording data (Unit 4)

Contents

Learning objectives

By the end of this topic you should be able to:

- *understand what you need to complete to pass the UASP associated with the project part of the course;*

- *be able to make an initial plan and make a timeline for the investigation;*

- *know how to look up ideas for suitable experiments and who to talk to about accessing and using equipment;*

- *understand how to take suitable pictures and draw relevant diagrams for your investigation.*

9.1   Introduction

As part of the Advanced Higher Physics course you will have to do an investigation on a suitable Physics topic and complete a written report at the end. This report will be marked by the SQA and is worth a total of 30 marks.

Internal assessment

The Investigation is a unit of the course and, as such, is also internally assessed using a UASP. The UASP has two outcomes which can be assessed from your record book by your teacher.

Outcome 1: Develop a plan for an investigation

Performance criteria.

a) A record is maintained in a regular manner.

b) Experimental and observational techniques and apparatus are appropriate for the investigation.

Outcome 2: Collect and analyse information obtained from the investigation

Performance criteria.

a) The collection of the experimental information is carried out with due accuracy.

b) Relevant measurements and observations are recorded in an appropriate format.

c) Recorded experimental information is analysed and presented in an appropriate format.

d) Uncertainties are treated appropriately.

Copyright ©Scottish Qualifications Authority
This may change from year to year, always check the SQA website for the most up to date mark scheme and assessment information.

9.2   The planning cycle and initial plan

The first stage in the process is to set up a timeline for the project such as one like this:

Phase	Start date	Tasks	Deadlines
Research and choose a topic	Now!	Check with teacher for suitability and equipment	
Experiment 1			
Experiment 2			
Experiment 3 (+4 etc.)			
Write report			
Hand in first draft			
Hand in final draft			

Secondly you need to research topics that you are interested in and perhaps trial some experiments to see if it will make a viable investigation. Research could be from the internet or from textbooks and journals. If you use books you should give the author, title, edition and page numbers and if you use the internet the full URL is needed. Both will need the date you looked at them too. The final report will need a minimum of two correctly referenced references mentioned to achieve the mark.

These are the correct way to record your references:

1. Research found from the internet - a full URL with the date accessed.

   http://www.cyberphysics.co.uk/topics/forces/young_modulus.htm - accessed on 10/10/2015

2. Research found from a textbook.

   Tom Duncan, A Textbook for Advanced Level Students, 2nd Edition, Pages 228 - 229. Read on 06/02/2015

You will have to check with your teacher whether equipment is available, sometimes a visit to a University can be arranged or equipment borrowed and most schools will have some equipment available. Some students even construct their own equipment for the investigation but bear in mind this will take up a lot of extra time.

Some ideas for experiments can be found here:

* http://www.sserc.org.uk/index.php/physics-home/advanced-higher/3362-investigation-ideas

* http://smarshallsay.weebly.com/advanced-higher.html

Don't be afraid to try something else that you may be interested in, but remember it needs to be at a level commensurate with AH and needs to have related experiments.

You will need to record all your research in your record book or "Day Book" which can be an actual experimental jotter, but could also be loose leaf paper in a folder, a file on a computer, etc. It should be checked and marked regularly by your teacher and don't forget they should contain all websites accessed or textbooks used and date accessed as well as all experimental data.

9.3 Using equipment and recording experimental data

Don't be surprised if the practical work takes many more hours than you planned, this is normal. Good planning will help here, making sure all equipment is ready in advance and your daybook is clearly laid out with dates and clearly drawn up tables of collected data. An account of the experimental procedures should also be written and written in the third person form.

Images

A **labelled** diagram of apparatus should be drawn (this can be in rough in the daybook) and take pictures of the apparatus too which can also be used in the final report. With pictures try and avoid background objects or overly crowded pictures which detract from the details in the image.

This image is a bit cluttered and not clear to the reader

"Bomb Calorimeter" by Akshat Goel, licensed under CC BY 3.0

Slightly better though still with a distracting background

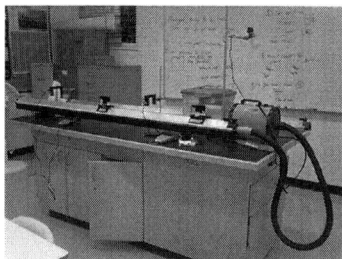

"Air Track with photo-gates and a reverse vacuum" by Bhavesh Chauhan is licensed under CC BY 3.0

Nice clear equipment with no distracting background which can be clearly labelled

"An Electrospinz Ltd Doris type laboratory electrospinning machine", by Robert Lamberts, licensed under CC BY 3.0

Diagrams

Similarly diagrams should be fully labelled and clear 2D diagrams not 3D works of art, try and think.

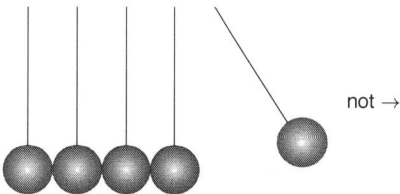

not →

Here are some examples of labelled diagrams that could be used in conjunction with labelled photos.

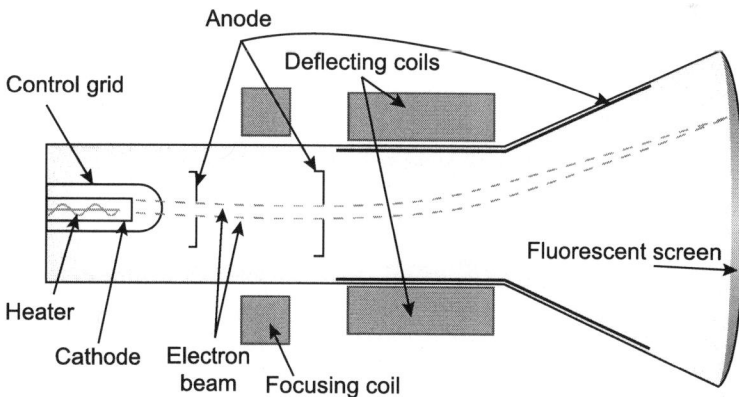

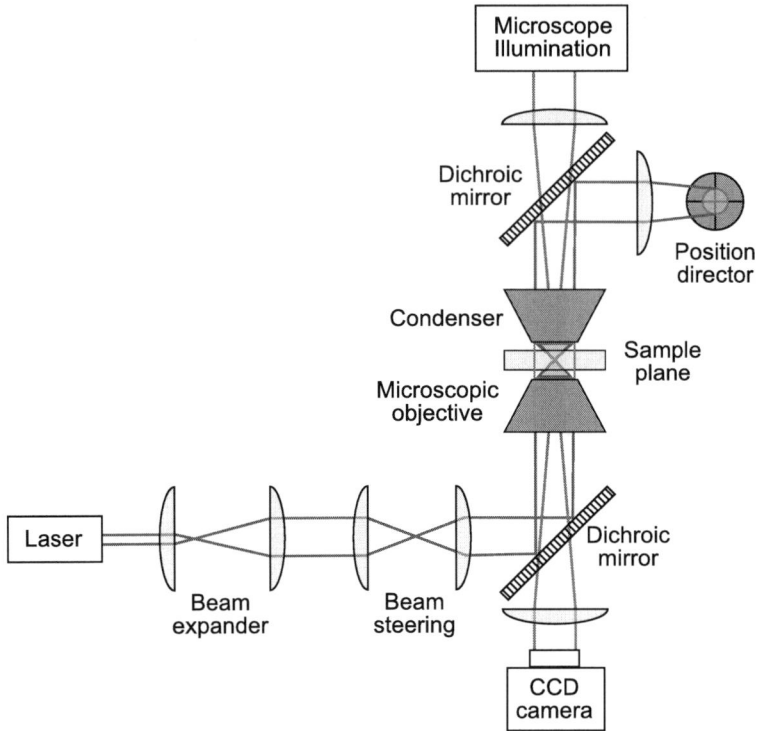

9.4 Summary

Summary

You should now be able to:

- understand what you need to complete to pass the UASP associated with the project part of the course;

- be able to make an initial plan and make a timeline for the investigation;

- know how to look up ideas for suitable experiments and who to talk to about accessing and using equipment;

- how to take suitable pictures and draw relevant diagrams for your investigation.

Topic 10

Measuring and presenting data (Unit 4)

Contents

Learning objectives

By the end of this topic you should be able to:

- *correctly complete results tables for your investigation;*

- *produce accurate graphs of the correct size and scale.*

10.1   Result tables

Don't forget that tables of data need to have correct units and headings. A rough graph is useful to identify outliers which may affect your results. You may wish to use a spreadsheet program such as Excel or Numbers to produce the graph, but make sure the axes are correctly labelled, the scale is suitable and there are enough gridlines to clearly see the data points. The data points should be suitably small and not obscure the grid lines and ideally should be error bars not dots. (See tips for drawing graphs below for more information.)

A best fit line or curve should be added to aid with analysis of data. Excel can also give you an equation of a best fit line. Uncertainties will need to be taken into account and they will be looked at in more detail in Unit 5. Also useful is to note down any difficulties encountered and how you dealt with them and any further improvements you can think of or further work you might like to look into. This will help greatly when writing up the evaluation part of the final report. Make sure you get plenty of data. When appropriate, a good scientist repeats readings enough times to calculate the random uncertainty and has a good range of data.

Example

Problem:

This is an example of a seemingly well presented table of results for a Newton's Second Law experiment involving varying the unbalanced force applied to various masses, note that this student has made a major mistake in their experiment and a few minor ones as well, can you spot their failings?

Mass (kg)	Force1 (N)	Force2 (N)	Force3 (N)	ForceAVG (N)	Acceleration
1.0	3.4	3.6	3.5	4	4
2.0	2.8	2.2	3.1	3	1.5
3.0	1.9	1.4	1.6	1.6	0.5
4.0	8.6	7.8	8.5	8	2.0
5.0	10.0	11.2	12.8	11.3	2.3
6.0	6.5	6.3	6.5	6.4	1.1
7.0	3.6	4.5	7.2	5	0.7
8.0	7.5	7.8	8.1	8	1.0
9.0	4.5	4.7	3.8	4.3	0.5

Things to note:

- Your results should have the correct number of significant figures/decimal places as appropriate. This will aid graph drawing later.

- You must have all headings labelled with the correct units.

- You may prefer to use a program such as Excel to perform the calculations in the columns for you, eg in this table the average force and acceleration were calculated using this method. This can save you a lot of time compared to using a calculator. See Excel Help if your teacher can't help you with this method.

Solution:

Mistakes in table:

- The student above has varied the mass **and** the unbalanced force making it difficult to plot a valid graph of results for their experimental data.

- Units for acceleration are missing.

- To calculate the random uncertainty in the Force, you would be advised to have two more repeats.

- Some of the Force and Acceleration values are only quoted to one significant figure.

. . . .

Tips for drawing graphs

Be careful when using computer programs to draw graphs for you, computers tend not to show minor grid lines, leaving you with floating points which are not at all accurate:

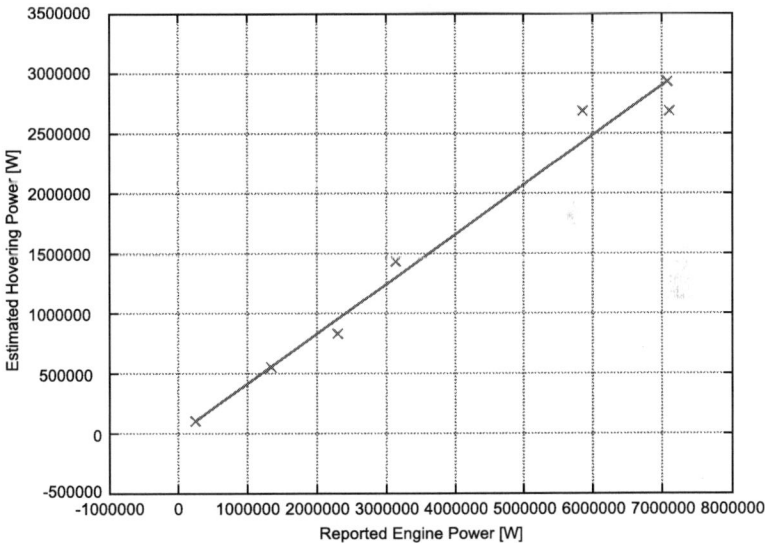

Another graph shown below is way too small on the page, the axes have not been labeled, the points are too large and so on.

Acceleration

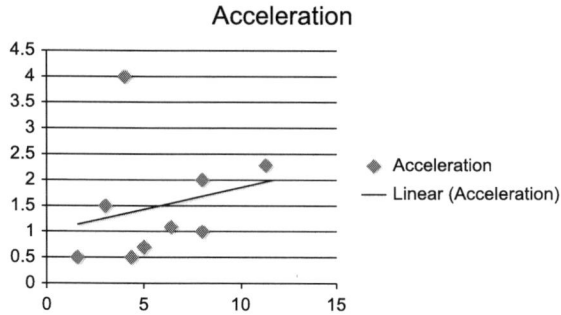

A much better version of the same graph would be:

Newton's Second Law Graph, Acceleration against Force

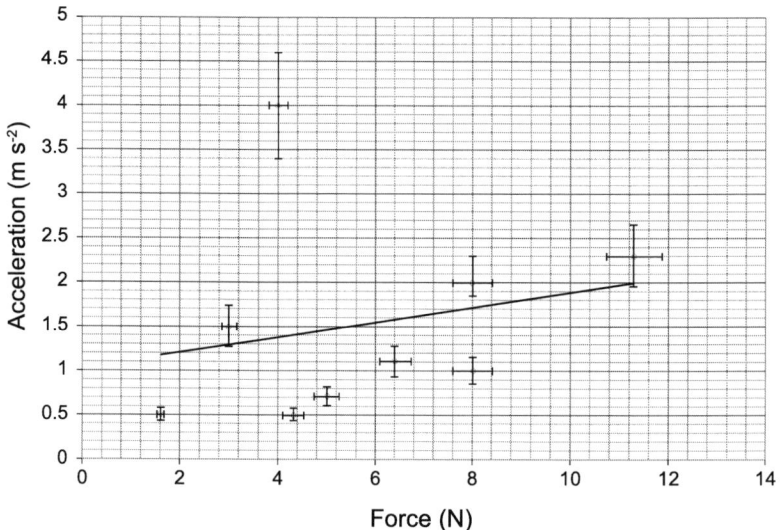

This graph has clearly labeled axes, major and minor gridlines and the large points have been replaced with error bars showing the uncertainties in the mass and acceleration. The point at 4N is an outlier and worthy of comment or even removal and the computer could also be used to calculate the gradient of the line for the conclusion. LINEST is a useful feature on excel for working out the uncertainty in the gradient if desired. Quite often a sharp pencil and a graph paper will enable you to be accurate enough if you prefer the paper and pencil method and the parallelogram method of uncertainties can work out the error in the gradient and intercept. (See Unit 5 for more information on this method).

10.2 Summary

Summary

You should now be able to:

- correctly complete results tables for your investigation;
- produce accurate graphs of the correct size and scale.

Topic 11

Evaluating findings (Unit 4)

Contents

Prerequisite knowledge

*It is recommended that you work through this section of **CfE Higher Physics** if you haven't already as these are the essential evaluative skills you need prior to reading this topic.*

- **Higher (CfE) physics - Unit 4 - Topic 5.4**

Learning objectives

By the end of this topic you should be able to:

- *evaluate your results and draw conclusions;*

- *understand some tips for completing the evaluation sections of the report.*

11.1   Evaluating findings

Evaluating your scientific findings - general tips.

This is a section that many candidates find it hard to pick up full marks and candidates are advised to use the following points to structure the discussion. Each experiment should have it's own conclusion and evaluation with a final overall evaluation at the end of the report.

Conclusions need to be valid and **related to the aim** of the investigation and they tend to take the form of the calculation of the gradient of a graph with appropriate uncertainties for example.

In the evaluation try and comment on all of the following:

1. Accuracy of experimental measurements

How did you ensure that each result was as accurate as it could be?

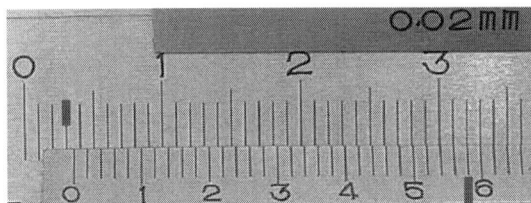

Zoom-in on "Messschieber.jpg" made by Ultraman,Wikimedia, licensed under CC BY 3.0

Using a Vernier scale instead of a regular ruler will vastly reduce your reading uncertainty.

Or being aware of effects such as PARALLAX, where the reading can change depending on how you look at it.

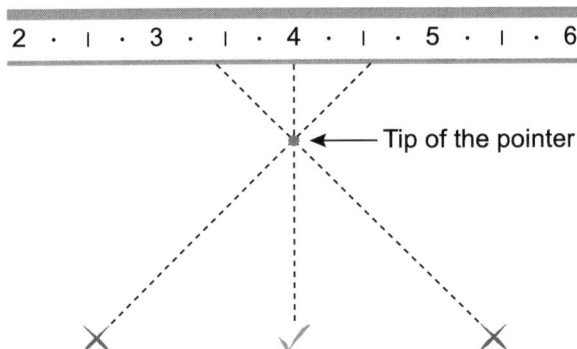

2. Adequacy of repeated readings

Did you have enough readings to calculate an accurate random uncertainty, was the equipment accurate enough to give you similar results each time?

3. Adequacy of range over which variables are altered

Quite often equipment restraints will prevent you having the range you desire, can you think of any ways to extend the range of your variables or at the least comment on these restrictions.

4. Adequacy of control of variables

How did you reduce uncertainties in each variable and how many fixed variables were there in your experiment and how did you control them?

5. Limitations of equipment

Most school equipment will have some failings, systematic uncertainties are quite common, look out and check for these. For example a Voltmeter that is always 0.2 mV out, a wooden ruler that is 0.5mm shorter than it once was, etc.

V-O-M (Volt-Ohm-Meter) / Multimeter, Author: Steve C, Source: Flickr

6. Reliability of methods

Be self-critical here, how reliable was your method and how could it be improved?

7. Sources of errors and uncertainties

Think about whether the uncertainties are most likely calibration, scale, reading etc or if human error is a major factor. Looking at the uncertainty calculations will make this much easier to target the largest sources of error.

The overall conclusion and evaluation of the investigation as a whole should mention:

- **Problems overcome** - During all three (or more) experiments.

    Examples:

    - A darkroom was needed to eliminate background light but one wasn't available.
    - The school multimeter/digital balance/data logger only has a limited range/large percentage uncertainty at smaller readings.
    - My school equipment wasn't advanced enough so I had to contact Scotland University for help with my project.
    - I had to limit the current to xA due to wires heating up.
    - etc.

- **Modifications to procedures and suggested future experiments** - If you had an unlimited budget what would you like to do to improve and enhance your experiments?

- **Significance/interpretation of findings** - Can you find any similar work to compare yours too, can you see a bigger picture from your findings? Can you link your results back to your background theory?

More detail of what to write in the evaluation can be found in the next Topic 5, which takes you through the process of writing up the scientific report.

11.2   Summary

Summary

You should now be able to:

- evaluate your results and draw conclusions;

- understand some tips for completing the evaluation sections of the report.

Topic 12

Scientific report (Unit 4)

Contents

Learning objectives

By the end of this topic you should be able to:

- *understand how to approach the project write up and will be familiar with the mark scheme used by the SQA.*

12.1 Scientific report

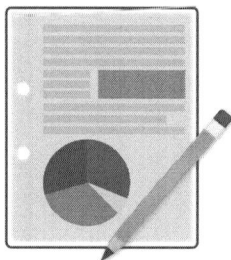

When you come to write up your project report, you need to make sure it is between 2000 and 3000 words in length. Too short and it will not be detailed enough to pick up some of the marks in the mark scheme and too long - over 3300 words (10% over the maximum) and you will suffer a penalty. The total mark is 30 marks and you should make sure you hand in drafts to your teacher well in advance of the submission date for checking and re-editing. The most successful candidates may hand in two or even three rough drafts before the final submission.

The best tip is to pay close attention to the mark scheme and tick off each section as you complete it. The following instructions are taken direct from the SQA website and give detailed instructions on how the marks are awarded with a few extra tips to make sure you get the best out of your report writing process.

Sections you should include in the report

1. Abstract   1 mark
2. Introduction   4 marks
3. Procedures   7 marks
4. Results   8 marks
5. Discussion   8 marks
6. Presentation   2 marks

Total = 30 marks

More detail on each section in the report

1. **Abstract (1 mark)**

   A brief abstract (summary) stating the overall aim(s) and finding(s)/conclusion(s) of the investigation.

   The abstract must be:

   - relevant to the investigation;
   - demonstrating an understanding of the physics theory underpinning the investigation;
   - of an appropriate Advanced Higher Physics level;
   - **immediately following the contents page** and should be under a separate heading. The abstract must be separate from and placed before the 'introduction';
   - the overall findings must be consistent with the conclusion(s) given in the 'discussion' and should relate to the aim(s).

2. **Introduction (4 marks)**

   - Candidates must include an account of the underlying physics that is relevant to the investigation. All terms and symbols used should be clearly defined. Simply stating equations is not sufficient - derivation of formulae should be given and all symbols in the equations must be explained with correct units. Candidates must demonstrate a good understanding of the physics behind these equations.
   - Candidates may (and should) draw on a variety of sources of information when researching their chosen topic. Don't base all your theory on one website of information.
   - Downloading directly from the internet or copying directly from books may suggest that the candidate has not understood the physics involved and will be considered as plagiarism. Where the vast majority is believed to have been copied verbatim then the candidate is not demonstrating understanding.
   - Complicated diagrams copied and pasted from an internet source are perfectly acceptable, especially when the reference is cited in the text and listed at the back of the report. Any form of referencing can be used but these should always be cited in the text.

3. **Procedures (7 marks)**

   Labelled diagrams and/or descriptions of apparatus (2 marks)

   - Candidates must include labelled diagrams and/or descriptions of the apparatus that they used for experimental work. Photographs of assembled apparatus, with appropriate labelling, are acceptable. A satisfactory photograph showing clear detail should be labelled as covered in Topic 3, and if possible use a labelled photograph and a labelled diagram for clarity.

   Clear descriptions of how the apparatus was used to obtain experimental readings (2 marks)

- Candidates must also give clear descriptions of how they used the apparatus to obtain their experimental results.
- The report should be written in the *past tense and impersonal voice*. (3rd person)
- Bulleted/numbered points are not recommended for the method. Use proper paragraphs to write your third person description.
- The procedure should be described well enough for another competent Advanced Higher Physics candidate to be able to repeat the procedure from the description.

Procedures are at an appropriate level for Advanced Higher (3 marks)

Factors to be considered include:

- range of procedures and number of repetitions;
- control of variables;
- accuracy;
- originality of approach and/or experimental techniques;
- degree of sophistication of experimental design and/or equipment.

Some of this is out of your hands due to equipment restraints but reading this list beforehand combined with talking to your teacher should help steer you in the correct direction to attain these marks. Use experiments completed in class as a rough guide to the standard required.

4. **Results (including uncertainties) (8 marks)**

Data sufficient and relevant to the aim(s) of the investigation (1 mark)

- The experimental data that candidates collect must be relevant to the aim(s) of their investigation. Also, the data candidates collect and present in their report must be sufficient in quantity and with a degree of accuracy and precision appropriate to their investigation - ie it must show all readings and not just the mean values. Don't forget too that data in tables is meaningless unless supplied with the correct labels and units.

Appropriate analysis of data, eg quality graphs, lines of best fit, calculations (4 marks)

- A candidate's report must include analysis of their experimental data that is appropriate to the investigation. This may involve drawing graphs or calculating and tabulating numerical values. Again further tables, graphs and calculations should still have the correct units and significant figures applied to them.

Uncertainties in individual and final results (3 marks)

- Candidates must include uncertainties in the values of each of the physical quantities that they measure and in the final result(s) of their investigation. Their analysis should show clearly how they have calculated/estimated the uncertainty in their final result(s). The best way to do this is with an example calculation for each method used, eg random uncertainty, percentage uncertainty, uncertainty in gradient of graph etc. Uncertainty calculations should all be at Advanced Higher level and all uncertainties (calibration, scale reading and random) that have a bearing on the accuracy on the experimental work should be mentioned. See Unit 5 for more information on uncertainties.

5. **Discussion (8 marks)**

Conclusion(s) is/are valid and relate to the aim(s) of the investigation (1 mark)

- Candidates must include overall conclusion(s) that are relevant to the aims(s) of their investigation and *supported by data* in the report and which are valid for the experimental results obtained.

Evaluation of experimental procedures (3 marks)

- Candidates must also include a critical evaluation of each experiment. It is often appropriate to include this after the 'procedures' and 'results' of each experiment. This should be a significant part of the candidate's report and should focus on the quality of their experimental work. See topic 4 for some more tips in writing this section. Candidates should include as many factors as possible and suggest improvements to procedures.
  - Accuracy of experimental measurements
  - Adequacy of repeated readings
  - Adequacy of range over which variables are altered
  - Adequacy of control of variables
  - Limitations of equipment
  - Reliability of methods
  - Sources of errors and uncertainties

  One tip here is to make sure you write something relevant for all seven points for each experiment being careful not to repeat yourself.

Coherent discussion of overall conclusion(s) and critical evaluation of the investigation as a whole (3 marks)

- Candidates must include a discussion of their overall conclusion(s) together with a critical evaluation of the investigation as a whole. This should be a more wide-ranging discussion of the investigation. It is an opportunity to explain what the candidate has learned as a result of the investigation and the significance of their findings. Candidates could also demonstrate the depth of their understanding of the physics related to the investigation.
  - Problems overcome
  - Modifications to procedures
  - Significance/interpretation of findings
  - Suggestions for further improvements to procedures
  - Suggestions for further work

Overall quality of the investigation (1 mark)

- This is a final quality mark for the standard of the investigation - not just the 'discussion' part of the report. This is for a good investigation well worked through, taking particular account of the physics involved and synthesis of argument.

6. **Presentation (2 marks)**

Appropriate structure, including informative title, contents page and page numbers (1 mark)

- The report structure should be easy to follow. A *title, contents page* and structure are essential - the contents page must show *page numbers* and the pages throughout the report must be numbered. Occasional missing page numbers (eg on hand-drawn graphs) will not be penalised. A title page with a nice picture or diagram to catch the marker's eye and show something of what the report is about is not essential but highly recommended.

References cited in the text and references listed in standard form, acknowledgements, where appropriate (1 mark)

- At *least three references* must be *cited correctly* in the main body of the report and the same ones also listed correctly at the back of the report. Any additional references cited or listed incorrectly should not be penalised. Any standard form of referencing is acceptable.
- References must be relevant to the investigation and specific. References must be cited within the text of the candidate's report and in many cases these will occur in the 'Underlying Physics' section. At the end of the report, the candidate must include details on all of the references (eg books, journals/periodicals and websites) that they cited. Candidates must include sufficient information to allow a reader to consult the original work to confirm its relevance to the investigation. Candidates should only include details on references; do not include information on materials that were part of background reading but are not cited as references in the report.

Total marks = 30 for the report

> **Top tip**
>
> Quite a lot of the advice above is **Copyright ©Scottish Qualifications Authority** and may change from year to year, always check the website for the most up to date mark scheme.

12.2 Summary

Summary

You should now be able to:

- have a thorough understanding of how to approach the project write up;

- be familiar with the mark scheme used by the SQA and know how to get as many of the 30 marks available as you can.

Glossary

Back e.m.f.

an induced e.m.f. in a circuit that opposes the current in the circuit.

Capacitive reactance

the opposition which a capacitor offers to current.

Conservative field

a field in which the work done in moving an object from one point in the field to another is independent of the path taken.

Coulomb's law

the electrostatic force between two point charges is proportional to the product of the two charges, and inversely proportional to the square of the distance between them.

Current-carrying conductor

exactly as its name suggests - a conductor of some sort in which there is a current.

Eddy currents

an induced current in any conductor placed in a changing magnetic field, or in any conductor moving through a fixed magnetic field.

Electric field

a region in which an electric charge experiences a Coulomb force.

Electric potential

the electric potential at a point in an electric field is the work done per unit positive charge in bringing a charged object from infinity to that point.

Electromagnetic braking

the use of the force generated by eddy currents to slow down a conductor moving in a magnetic field.

Faraday's law of electromagnetic induction

the magnitude of an e.m.f. produced by electromagnetic induction is proportional to the rate of change of magnetic flux through the coil or circuit.

Ferromagnetic

materials in which the magnetic fields of the atoms line up parallel to each other in regions known as magnetic domains.

Fundamental unit of charge

the smallest unit of charge that a particle can carry, equal to 1.60×10^{-19} C.

High-pass filter

an electrical filter that allows high frequency signals to pass, but blocks low frequency signals.

Induced e.m.f.

the e.m.f. induced in a conductor by electromagnetic induction.

Induction heating

the heating of a conductor because of the eddy currents within it.

Inductive reactance

the opposition which an inductor offers to current.

Inductor

acoil that generates an e.m.f. by self-inductance. The inductance of an inductor is measured in henrys (H).

Lenz's law

the induced current produced by electromagnetic induction is always in such a direction as to oppose the change that is causing it.

Magnetic domains

regions in a ferromagnetic material where the atoms are aligned with their magnetic fields parallel to each other.

Magnetic flux

a measure of the quantity of magnetism in a given area. Measured in weber (Wb), equivalent to $T\ m^2$.

Magnetic induction

a means of quantifying a magnetic field.

Magnetic poles

one way of describing the magnetic effect, especially with permanent magnets. There are two types of magnetic poles - north and south. Opposite poles attract, like poles repel.

Permeability of free space

a constant used in electromagnetism. It has the symbol μ_0 and a value of $4\pi \times 10^{-7}\ H\ m^{-1}$ (or $T\ m\ A^{-1}$).

Potential difference

the potential difference between two points is the difference in electric potential between the points. Since electric potential tells us how much work is done in moving a positive charge from infinity to a point, the potential difference is the work done in moving unit positive charge between two points. Like electric potential, potential difference V is measured in volts V.

Principle of superposition of forces

the total force acting on an object is equal to the vector sum of all the forces acting on the object.

Self-inductance

the generation of an e.m.f. by electromagnetic induction in a coil owing to the current in the coil.

Strong nuclear force

the force that acts between nucleons (protons and neutrons) in a nucleus, binding the nucleus together.

Time constant

the time taken for the charge stored by a capacitor to increase by 63% of the difference between initial charge and full charge, or the time taken to discharge a capacitor to 37% of the initial charge.

Weak nuclear force

a nuclear force that acts on particles that are not affected by the strong force.

Hints for activities

Topic 1: Electric force and field (Unit 3)

Quiz: Coulomb force

Hint 1: Remember Newton's Third law.

Hint 2: The number of electrons in 1 C is equal to the inverse of the fundamental charge.

Hint 3: This is a straight application of Coulomb's Law.

Hint 4: This is a straight application of Coulomb's Law.

Hint 5: Work out the size and direction of the force exerted by X on Y. Then work out the size and direction of the force exerted by Z on Y. Then add the two vectors.

Quiz: Electric field

Hint 1: How does the strength of the electric force exerted by a point charge vary with distance?

Hint 2: Electric field strength is the force per unit positive charge.

Hint 3: Electric field strength is the force per unit positive charge.

Hint 4: Work out the size and direction of the electric field due to the 30 nC. Then work out the size and direction of the electric field due to the 50 nC. Then add the two vectors.

Hint 5: Electric field is zero at the point where the magnitude of the field due to the 1.0 μC charge is equal to the magnitude of the field due to the 4.0 μC charge.

Topic 2: Electric potential (Unit 3)

Quiz: Potential and electric field

Hint 1: This is a straight application of $V = Ed$.

Hint 2: Make E the subject of the relationship $V = Ed$; then consider units on both sides of the equation.

Hint 3: This is a straight application of $E_W = QV$.

Hint 4: This is a straight application of $E_W = QV$.

Quiz: Electrical potential due to point charges

Hint 1: This is a straight application of $V = \frac{Q}{4\pi\varepsilon_0 r}$.

Hint 2: The charge of an alpha particle is double the charge of an electron. Use $E_p = E_W = QV$.

Hint 3: Find the potential due to each charge using $V = \frac{Q}{4\pi\varepsilon_0 r}$. Don't forget to include the minus sign for negative charges here.

Hint 4: To find out how $\frac{E}{V}$ depends on r, substitute $E = \frac{Q}{4\pi\varepsilon_0 r^2}$ and $V = \frac{Q}{4\pi\varepsilon_0 r}$ in $\frac{E}{V}$.

Topic 3: Motion in an electric field (Unit 3)

Quiz: Acceleration and energy change

Hint 1: Is the velocity of the electron increasing, decreasing or staying the same?

Hint 2: This is a straight application of $E_W = QV$.

Hint 3: The electrical energy QV is converted to kinetic energy $\frac{1}{2}mv^2$.

Hint 4: Use $V = Ed$ and then $E_W = QV$.

Quiz: Charged particles moving in electric fields

Hint 1: Electric field strength is the force per unit (positive) charge.

Hint 2: First find the electrical force, then use Newton's Second Law.

Hint 3: Electric field strength is the force per unit **positive** charge.

Hint 4: Electric field strength is the force per **unit** positive charge.

Hint 5: What is the initial value of the vertical velocity of the electron? Find the vertical electrical force and use this to calculate the vertical acceleration of the electron. Then use the first equation of motion.

Topic 4: Magnetic fields (Unit 3)

Quiz: Magnetic fields and forces

Hint 1: See the section titled Magnetic forces and fields.

Hint 2: See the section titled Magnetic forces and fields.

Hint 3: See the section titled Magnetic forces and fields.

Quiz: Current-carrying conductors

Hint 1: See the section titled Force on a current-carrying conductor in a magnetic field.

Hint 2: See the section titled Force on a current-carrying conductor in a magnetic field.

Hint 3: See the section titled Magnetic Field Around a current-carrying conductor.

Hint 4: See the section titled Magnetic induction.

The hiker

Hint 1: (a) What is the expression for the magnetic field due to a current-carrying conductor?

Hint 2: (b) What is the maximum value of B at the new position?

Hint 3: (b)

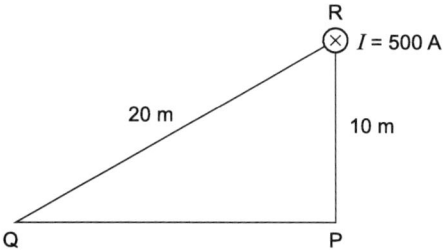

What is the horizontal distance PQ in relation to QR and PR?

Topic 6: Inductors (Unit 3)

Quiz: Self-inductance

Hint 1: This is a straight application of $\varepsilon = -L\frac{dI}{dt}$.

Hint 2: What is the rate of change of current?

Hint 3: This is an application of Lenz's law.

Hint 4: This is a straight application of $E = \frac{1}{2}LI^2$

Hint 5: This is a straight application of $E = \frac{1}{2}LI^2$

Quiz: Inductors in d.c. circuits

Hint 1: The maximum current is the steady value reached when the induced e.m.f. is zero.

Hint 2: The maximum potential difference across the inductor is the value when current in the circuit is zero.

Hint 3: The current is steady!!

Hint 4: For 'growth' read 'variation of current with time'.

Quiz: a.c. circuits

Hint 1: $X_L = 2\pi fL$.

Hint 2: See the section titled Inductors in a.c. circuits.

Hint 3: $X_L = 2\pi f L.$

Answers to questions and activities

1 Electric force and field (Unit 3)

Three charged particles in a line (page 7)

Expected answer

force

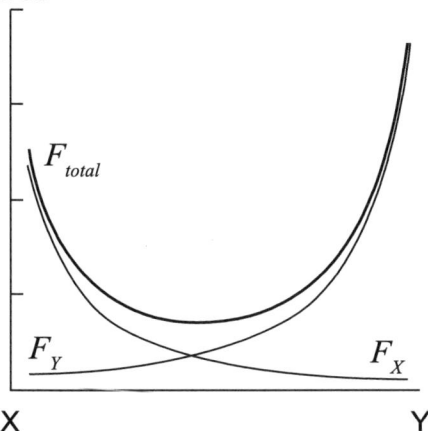

The graph shows the forces on the third charge due to charges X and Y, and the total force. As the third charge is moved from X to Y, the magnitude of the force due to X decreases whilst the magnitude of the force due to Y increases. The two forces both act in the **same** direction.

Quiz: Coulomb force (page 8)

Q1: c) F N

Q2: d) 6.25×10^{18}

Q3: d) 9.0 N towards B.

Q4: a) 7.7 cm

Q5: b) 14.4 N towards Z

Quiz: Electric field (page 14)

Q6: a) $E/4$

Q7: e) 1.25×10^{-8} N

Q8: d) 8.00×10^{-18} N in the -x-direction

Q9: d) 180 N C^{-1}

Q10: b) 17 cm

End of topic 1 test (page 18)

Q11: 2.0×10^{17}

Q12: 3.06 N

Q13: 9.2 N

Q14: 25 N

Q15: 5.7 N C^{-1}

Q16: 1.5×10^3 N C^{-1}

Q17: 3.2×10^5 N C^{-1}

Q18: 2.3×10^9 m s^{-2}

2 Electric potential (Unit 3)

Quiz: Potential and electric field (page 24)

Q1: c) 8.00 V

Q2: d) V m^{-1}

Q3: a) 0.24 J

Q4: e) 8000 V

Quiz: Electrical potential due to point charges (page 30)

Q5: c) 1.4×10^5 V

Q6: e) 4.48×10^{-14} J

Q7: a) -7.2×10^4 V

Q8: b) 50 m^{-1}

End of topic 2 test (page 32)

Q9: 270 V

Q10: 0.063 J

Q11: 0.050 m

Q12: 26 J

Q13: 3.8×10^5 V

Q14: 59 V

Q15: 1.35×10^4 V

Q16:

1. 2.3 m
2. 1.6×10^{-6} C

3 Motion in an electric field (Unit 3)

Quiz: Acceleration and energy change (page 38)

Q1: a) The electron gains kinetic energy.

Q2: b) 1.2×10^{-4} J

Q3: c) 3.10×10^5 m s^{-1}

Q4: b) 9.6×10^{-17} J

Quiz: Charged particles moving in electric fields (page 42)

Q5: c) 4.00×10^{-16} N

Q6: d) 7.03×10^{14} m s^{-2}

Q7: a) accelerated in the direction of the electric field.

Q8: c) 2.41×10^9 m s^{-2} downwards

Q9: d) 3.51×10^6 m s^{-1}

Rutherford scattering (page 49)

Expected answer

Use the formula

$$E_W = QV$$
$$\therefore E_W = eV_{gold}$$
$$\therefore E_W = e \times \frac{Q_{gold}}{4\pi\varepsilon_0 r}$$
$$\therefore r = e \times \frac{79e}{4\pi\varepsilon_0 E_W}$$

Now put in the values given in the question

$$r = \frac{79e^2}{4\pi\varepsilon_0 \times 8.35 \times 10^{-14}}$$
$$\therefore r = \frac{2.02 \times 10^{-36}}{9.29 \times 10^{-24}}$$
$$\therefore r = 2.18 \times 10^{-13} \text{ m}$$

End of topic 3 test (page 50)

Q10: 4.48×10^{-17} J

Q11: 5.1×10^6 m s^{-1}

Q12: 2.0×10^4 m s^{-2}

Q13:

1. 3.63×10^5 m s^{-1}
2. 2.34×10^{-3} m

Q14: 6.67×10^{-14} J

4 Magnetic fields (Unit 3)

Quiz: Magnetic fields and forces (page 58)

Q1:   e) All magnets have two poles called north and south.

Q2:   e) (i) and (iii) only

Q3:   c) (iii) only

Oersted's experiment (page 59)

Expected answer

1. A current through the wire produces a circular magnetic field centred on the wire.

2. The greater the current, the stronger is the magnetic field. This is shown by the separation of the field lines.

3. If the direction of the current is reversed, the direction of the magnetic field is also reversed.

Quiz: Current-carrying conductors (page 70)

Q4:   e) field the same, current doubled, length halved

Q5:   d) 60°

Q6:   e) circular, decreasing in magnitude with distance from the wire

Q7:   b) N A^{-1} m^{-1}

The hiker (page 72)

Expected answer

a)
$$B = \frac{\mu_0 I}{2\pi r}$$
$$\therefore B = \frac{4\pi \times 10^{-7} \times 500}{2\pi \times 10}$$
$$\therefore B = 1.0 \times 10^{-5} \text{ T}$$

b)

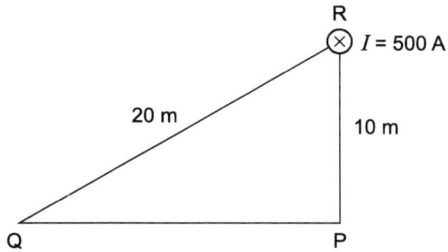

$$B_{\text{Earth}} = 0.5 \times 10^{-4} \text{ T}$$
$$\therefore 10\% \text{ of } B_{\text{Earth}} = 0.5 \times 10^{-5} \text{ T}$$

$$B = \frac{\mu_0 I}{2\pi r}$$
$$\therefore 0.5 \times 10^{-5} = \frac{4\pi \times 10^{-7} \times 500}{2\pi \times \text{QR}}$$
$$\therefore \text{QR} = \frac{4\pi \times 10^{-7} \times 500}{2\pi \times 0.5 \times 10^{-5}}$$
$$\therefore \text{QR} = 20 \text{ m}$$

$$\text{QP} = \sqrt{20^2 - 10^2}$$
$$\therefore \text{QP} = \sqrt{400 - 100}$$
$$\therefore \text{QP} = \sqrt{300}$$
$$\therefore \text{QP} = 17.3 \text{ m}$$

Electrostatic and gravitational forces (page 74)

Expected answer

The Coulomb force is given by

$$F_C = \frac{Q_1 Q_2}{4\pi\varepsilon_0 r^2}$$
$$\therefore F_C = \frac{\left(1.6 \times 10^{-19}\right)^2}{4\pi \times 8.85 \times 10^{-12} \times \left(10^{-15}\right)^2}$$
$$\therefore F_C \sim \frac{2.5 \times 10^{-38}}{1 \times 10^{-40}}$$
$$\therefore F_C \sim 250 \text{ N}$$

The gravitational force is given by

$$F_G = G\frac{m_1 m_2}{r^2}$$

$$\therefore F_G = 6.67 \times 10^{-11} \times \frac{\left(1.67 \times 10^{-27}\right)^2}{\left(10^{-15}\right)^2}$$

$$\therefore F_G \sim 6.67 \times 10^{-11} \times \frac{2.8 \times 10^{-54}}{10^{-30}}$$

$$\therefore F_G \sim 1.9 \times 10^{-34}\text{N}$$

We can combine these two results to find the ratio F_C/F_G

$$F_C/F_G \sim \frac{250}{1.9 \times 10^{-34}}$$

$$\therefore F_C/F_G \sim 10^{36}$$

End of topic 4 test (page 79)

Q8: 0.0254 N

Q9: 276 N

Q10: 4.37 m s^{-2}

Q11:

  1. b) added to QR.
  2. 0.21 T

Q12: 2.3 A

Q13: 4.03 $\times$ 10^{-5} T

Q14:

  1. 1.7 $\times$ 10^{-3} T
  2. 4.5 $\times$ 10^{3} A

5 Capacitors (Unit 3)

End of topic 5 test (page 103)

Q1: 45 s

Q2: 150 Ω

Q3: 51 mA

Q4: d)

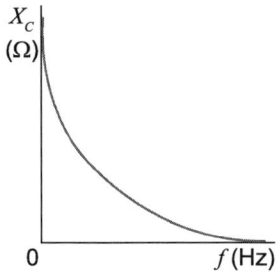

Q5:

1. 152 Ω
2. 0.092 A

6 Inductors (Unit 3)

Quiz: Self-inductance (page 119)

Q1: b) moved across a magnetic field.

Q2: d) the induced current is always in such a direction as to oppose the change that is causing it.

Q3: b) 0.29 H

Q4: a) 0 V

Q5: d) The self-induced e.m.f. in an inductor always opposes the change in current that is causing it.

Q6: a) 0.18 J

Q7: d) 0.85 H

Quiz: Inductors in d.c. circuits (page 124)

Q8: b) 25 mA

Q9: d) 1.5 V

Q10: a) 0 V

Q11: d) stopwatch

Quiz: a.c. circuits (page 128)

Q12: c) $X \propto f$

Q13: e) An inductor can smooth a signal by filtering out high frequency noise and spikes.

Q14: a) $I \propto {}^1\!/_f$

End of topic 6 test (page 131)

Q15: 8.3 V

Q16: 0.48 J

Q17: c) 1 V s A^{-1}

Q18: 0.048 V

Q19: 8.2 V

Q20: 8.4 V

Q21:

1. 2.6 V
2. 0 V
3. 0.19 A

Q22:

1. 0.65 V
2. 0.043 A
3. 3.1×10^{-4} J

Q23:

1. 0.12 H
2. 0.18 A
3. 1.9×10^{-3} J

Q24:

1. 2.8 V
2. 2.8 V
3. 4.9×10^{-3} J

7 Electromagnetic radiation (Unit 3)

End of topic 7 test (page 141)

Q1: ε_0 is the symbol for the **permittivity** of free space.

Q2:   Electromagnetic waves are **transverse**.

Q3: Electricity and magnetism can be **unified** under one theory called electromagnetism.

Q4:   d) $c = \frac{1}{\sqrt{\varepsilon_0 \mu_0}}$

Q5: 1.44×10^{-6} H m^{-1}

8 End of unit 3 test

End of unit 3 test (page 144)

Q1:

1. 2.00×10^5 N

2. 3.28×10^4 N

Q2: 6.3×10^{-6} T

Q3: 5.1×10^{-26} kg

Q4:   4.1 N

Q5:

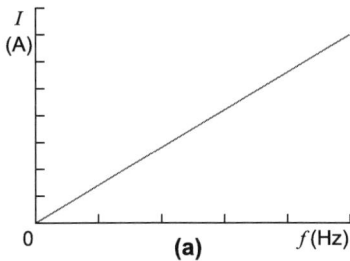

Q6:   170 Ω

Q7:

1. 21 A s^{-1}

2. 3.9×10^{-3} J

Q8:   7.1 V

Q9:

1. 3.2 V

2. 0 V

3. 0.23 A

Q10:

1. 0.133 H

2. 0.200 A

3. 2.67×10^{-3} J

Q11:

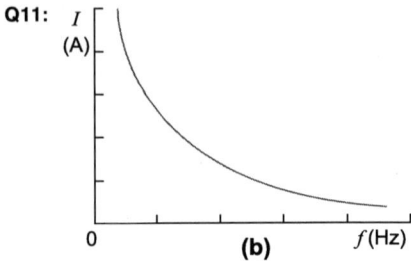

(b)

Q12:

time = 0.075 s